Albert Einstein

RELATIVITY

Edited by Vesselin Petkov

MINKOWSKI
Institute Press

Albert Einstein
Born on March 14, 1879 in Ulm, Germany
Died on April 18, 1955 in Princeton, New Jersey, United States

© Minkowski Institute Press
Published: 2018

ISBN: 978-1-927763-68-1 (softcover)
ISBN: 978-1-927763-69-8 (ebook)

Minkowski Institute Press
Montreal, Quebec, Canada
http://minkowskiinstitute.org/mip/

For information on all Minkowski Institute Press publications visit our web-
site at http://minkowskiinstitute.org/mip/books/

Preface by the Editor

The main part of this volume is a new publication of Einstein's valuable book *Relativity: The Special and the General Theory*, written for a wider audience. Four other articles, which nicely complement the issues discussed in the book, are also included in the volume.

The five works by Einstein included in this volume are:

1. *Relativity: The Special and the General Theory. A Popular Exposition*, translated by R. W. Lawson (Methuen And Co. Limited 1920)

2. "What is the Theory of Relativity," written at the request of *The London Times* and published on November 28, 1919, translated by R. W. Lawson

3. "Ether and the Theory of Relativity," An Address delivered on May 5th, 1920 in the University of Leyden, in: *Sidelights on relativity: I. Ether and Relativity, II. Geometry and Experience*, translated by G. B. Jeffery and W. Perrett (Methuen And Co. Limited 1922) pp. 1-24

4. "Geometry and Experience," An expanded form of an Address to the Prussian Academy of Sciences in Berlin on January 27th, 1921, in: *Sidelights on relativity: I. Ether and Relativity, II. Geometry and Experience*, translated by G. B. Jeffery and W. Perrett (Methuen And Co. Limited 1922) pp. 25-56

5. "A Brief Outline of the Development of the Theory of Relativity," *Nature*, volume 106, pages 782-784 (17 February 1921), translated by R. W. Lawson

One notable issue – the concept of relativistic mass – is not discussed in Einstein's five works republished here. Although he wrote in his popular book "the inertial mass of a body is not a constant" (p. 36), Einstein did not discuss the concept of relativistic mass (i.e., the relativistic increase of the mass of a body as its velocity increases), which he silently abandoned after using it in his 1905 paper, where he defined longitudinal mass and transverse mass. As Einstein's unclear view of the relativistic mass might have provided some encouragement for "what has probably been the most vigorous campaign ever waged against the concept of relativistic mass" (M. Jammer, *Concepts of Mass in Contemporary Physics and Philosophy* (Princeton University Press, Princeton 2000) p. 51), the Appendix *On Relativistic Mass* is included in this volume (p. 117) to address that unprecedented campaign and to help the readers make their own decision on the concept of relativistic mass and on that campaign.

9 September 2018   Vesselin Petkov
Montreal   *Minkowski Institute*

Preface by the Editor

Contents

Preface by the Editor   i

RELATIVITY: THE SPECIAL AND THE GENERAL THE-
ORY   1

Preface   2

I   I. The Special Theory of Relativity   7

1   Physical Meaning of Geometrical Propositions   9

2   The System of Coordinates   11

3   Space and Time in Classical Mechanics   13

4   The Galileian System of Coordinates   15

5   The Principle of Relativity (In the Restricted Sense)   16

6   The Theorem of the Addition of Velocities employed in
Classical Mechanics   18

7   The Apparent Incompatibility of the Law of Propagation
of Light with the Principle of Relativity   19

8   On the Idea of Time in Physics   21

9   The Relativity of Simultaneity   23

10 On the Relativity of the Conception of Distance   25

11 The Lorentz Transformation   26

12 The Behaviour of Measuring-Rods and Clocks in Motion   29

13 Theorem of the Addition of Velocities. The Experiment of Fizeau   31

14 The Heuristic Value of the Theory of Relativity   34

15 General Results of the Theory   35

16 Experience and the Special Theory of Relativity   39

17 Minkowski's Four-dimensional Space   42

II   II. The General Theory of Relativity   45

18 Special and General Principle of Relativity   47

19 The Gravitational Field   50

20 The Equality of Inertial and Gravitational Mass as an Argument for the General Postulate of Relativity   52

21 In what Respects are the Foundations of Classical Mechanics and of the Special Theory of Relativity unsatisfactory?   55

22 A Few Inferences from the General Principle of Relativity   57

23 Behaviour of Clocks and Measuring-Rods on a Rotating Body of Reference   60

24 Euclidean and Non-Euclidean Continuum   63

25 Gaussian Coordinates   65

26 The Space-Time Continuum of the Special Theory of Relativity considered as an Euclidean Continuum   68

27 The Space-Time Continuum of the General Theory of Relativity is not an Euclidean Continuum   70

28 Exact Formulation of the General Principle of Relativity   72

29 The Solution of the Problem of Gravitation on the Basis of the General Principle of Relativity   74

III   III. Consideration on the Universe as a Whole   77

30 Cosmological Difficulties of Newton's Theory   79

31 The Possibility of a "Finite" and yet "Unbounded" Universe   81

32 The Structure of Space according to the General Theory of Relativity   84

APPENDICES   87

1 Simple Derivation of the Lorentz Transformation   87
2 Minkowski's Four-Dimensional Space ("World")   91
3 The Experimental Confirmation of the General Theory of Relativity 93
4 The Structure of Space According to the General Theory of Relativity 99
5 Relativity and the Problem of Space 101

Bibliography   115

APPENDIX (by the Editor): On Relativistic Mass   117

WHAT IS THE THEORY OF RELATIVITY   127

ETHER AND THE THEORY OF RELATIVITY   131

GEOMETRY AND EXPERIENCE   139

A BRIEF OUTLINE OF THE DEVELOPMENT OF THE THEORY OF RELATIVITY   149

Notable Quotes by Einstein   155

Index   161

RELATIVITY: THE SPECIAL AND THE GENERAL THEORY

A POPULAR EXPOSITION

AUTHORISED TRANSLATION BY ROBERT W. LAWSON

PREFACE

The present book is intended, as far as possible, to give an exact insight into the theory of Relativity to those readers who, from a general scientific and philosophical point of view, are interested in the theory, but who are not conversant with the mathematical apparatus of theoretical physics. The work presumes a standard of education corresponding to that of a university matriculation examination, and, despite the shortness of the book, a fair amount of patience and force of will on the part of the reader. The author has spared himself no pains in his endeavour to present the main ideas in the simplest and most intelligible form, and on the whole, in the sequence and connection in which they actually originated. In the interest of clearness, it appeared to me inevitable that I should repeat myself frequently, without paying the slightest attention to the elegance of the presentation. I adhered scrupulously to the precept of that brilliant theoretical physicist L. Boltzmann, according to whom matters of elegance ought to be left to the tailor and to the cobbler. I make no pretence of having withheld from the reader difficulties which are inherent to the subject. On the other hand, I have purposely treated the empirical physical foundations of the theory in a "step-motherly" fashion, so that readers unfamiliar with physics may not feel like the wanderer who was unable to see the forest for trees. May the book bring some one a few happy hours of suggestive thought!

December, 1916   A. Einstein

NOTE TO THE THIRD EDITION

In the present year (1918) an excellent and detailed manual on the general theory of relativity, written by H. Weyl, was published by the firm Julius Springer (Berlin). This book, entitled *Raum—Zeit—Materie* (Space—Time—Matter), may be warmly recommended to mathematicians and physicists.

TRANSLATOR'S NOTE

In presenting this translation to the English-reading public, it is hardly necessary for me to enlarge on the Author's prefatory remarks, except to draw attention to those additions to the book which do not appear in the original.

At my request, Professor Einstein kindly supplied me with a portrait of himself, by one of Germany's most celebrated artists. Appendix 3 on "The Experimental Confirmation of the General Theory of Relativity," has been written specially for this translation. Apart from these valuable additions to the book, I have included a biographical note on the Author, and, at the end of the book, an Index and a list of English references to the subject. This list, which is more suggestive than exhaustive, is intended as a guide to those readers who wish to pursue the subject farther.

I desire to tender my best thanks to my colleagues Professor S. R. Milner, D.Sc., and Mr. W. E. Curtis, A.R.C.Sc., F.R.A.S., also to my friend Dr. Arthur Holmes, A.R.C.Sc., F.G.S., of the Imperial College, for their kindness in reading through the manuscript, for helpful criticism, and for numerous suggestions. I owe an expression of thanks also to Messrs. Methuen for their ready counsel and advice, and for the care they have bestowed on the work during the course of its publication.

Robert W. Lawson
The Physics Laboratory
The University of Sheffield
June 12, 1920

BIOGRAPHICAL NOTE

Albert Einstein is the son of German-Jewish parents. He was born in 1879 in the town of Ulm, Würtemberg, Germany. His schooldays were spent in Munich, where he attended the *Gymnasium* until his sixteenth year. After leaving school at Munich, he accompanied his parents to Milan, whence he proceeded to Switzerland six months later to continue his studies.

From 1896 to 1900 Albert Einstein studied mathematics and physics at the Technical High School in Zurich, as he intended becoming a secondary school (*Gymnasium*) teacher. For some time afterwards he was a private tutor, and having meanwhile become naturalised, he obtained a post as engineer in the Swiss Patent Office in 1902 which position he occupied till 1909. The main ideas involved in the most important of Einstein's theories date back to this period. Amongst these may be mentioned: *The Special Theory of Relativity, Inertia of Energy, Theory of the Brownian Movement,* and the *Quantum-Law of the Emission and Absorption of Light* (1905). These were followed some years later by the *Theory of the Specific Heat of Solid Bodies,* and the fundamental idea of the *General Theory of Relativity.*

During the interval 1909 to 1911 he occupied the post of Professor *Extraordinarius* at the University of Zurich, afterwards being appointed to the University of Prague, Bohemia, where he remained as Professor *Ordinarius* until 1912. In the latter year Professor Einstein accepted a similar chair at the *Polytechnikum*, Zurich, and continued his activities there until 1914, when he received a call to the Prussian Academy of Science, Berlin, as successor to Van't Hoff. Professor Einstein is able to devote himself freely to his studies at the Berlin Academy, and it was here that he succeeded in completing his work on the *General Theory of Relativity* (1915–17). Professor Einstein also lectures on various special branches of physics at the University of Berlin, and, in addition, he is Director of the Institute for Physical Research of the *Kaiser Wilhelm Gesellschaft*.

Professor Einstein has been twice married. His first wife, whom he married at Berne in 1903, was a fellow-student from Serbia. There were two sons of this marriage, both of whom are living in Zurich, the elder being sixteen years of age. Recently Professor Einstein married a widowed cousin, with whom he is now living in Berlin.

R. W. L.

NOTE TO THE FIFTEENTH EDITION

In this edition I have added, as a fifth appendix, a presentation of my views on the problem of space in general and on the gradual modifications of our ideas on space resulting from the influence of the relativistic viewpoint. I wished to show that space-time is not necessarily something to which one can ascribe a separate existence, independently of the actual objects of physical reality. Physical objects are not *in space*, but these objects are *spatially extended*. In this way the concept "empty space" loses its meaning.

June 9th 1952   A. Einstein

I. THE SPECIAL THEORY OF RELATIVITY

1 PHYSICAL MEANING OF GEOMETRICAL PROPOSITIONS

In your schooldays most of you who read this book made acquaintance with the noble building of Euclid's geometry, and you remember—perhaps with more respect than love—the magnificent structure, on the lofty staircase of which you were chased about for uncounted hours by conscientious teachers. By reason of your past experience, you would certainly regard everyone with disdain who should pronounce even the most out-of-the-way proposition of this science to be untrue. But perhaps this feeling of proud certainty would leave you immediately if some one were to ask you: "What, then, do you mean by the assertion that these propositions are true?" Let us proceed to give this question a little consideration.

Geometry sets out from certain conceptions such as "plane," "point," and "straight line," with which we are able to associate more or less definite ideas, and from certain simple propositions (axioms) which, in virtue of these ideas, we are inclined to accept as "true." Then, on the basis of a logical process, the justification of which we feel ourselves compelled to admit, all remaining propositions are shown to follow from those axioms, i.e., they are proven. A proposition is then correct ("true") when it has been derived in the recognised manner from the axioms. The question of the "truth" of the individual geometrical propositions is thus reduced to one of the "truth" of the axioms. Now it has long been known that the last question is not only unanswerable by the methods of geometry, but that it is in itself entirely without meaning. We cannot ask whether it is true that only one straight line goes through two points. We can only say that Euclidean geometry deals with things called "straight lines," to each of which is ascribed the property of being uniquely determined by two points situated on it. The concept "true" does not tally with the assertions of pure geometry, because by the word "true" we are eventually in the habit of designating always the correspondence with a "real" object; geometry, however, is not concerned with the relation of the ideas involved in it to objects of experience, but only with the logical connection of these ideas among themselves.

It is not difficult to understand why, in spite of this, we feel constrained to call the propositions of geometry "true." Geometrical ideas correspond to more or less exact objects in nature, and these last are undoubtedly the exclusive cause of the genesis of those ideas. Geometry ought to refrain from such a course, in order to give to its structure the largest possible logical unity. The practice, for example, of seeing in a "distance" two marked positions on a practically rigid body is something which is lodged deeply in our habit of thought. We are accustomed further to regard three points as being situated on a straight line, if their apparent positions can be made to coincide for observation with one eye, under suitable choice of our place of observation.

If, in pursuance of our habit of thought, we now supplement the propositions of Euclidean geometry by the single proposition that two points on a practically rigid body always correspond to the same distance (line-interval), independently of any changes in position to which we may subject the body, the propositions of Euclidean geometry then resolve themselves into propositions on the possible relative position of practically rigid bodies.[1] Geometry which has been supplemented in this way is then to be treated as a branch of physics. We can now legitimately ask as to the "truth" of geometrical propositions interpreted in this way, since we are justified in asking whether these propositions are satisfied for those real things we have associated with the geometrical ideas. In less exact terms we can express this by saying that by the "truth" of a geometrical proposition in this sense we understand its validity for a construction with ruler and compasses.

Of course the conviction of the "truth" of geometrical propositions in this sense is founded exclusively on rather incomplete experience. For the present we shall assume the "truth" of the geometrical propositions, then at a later stage (in the general theory of relativity) we shall see that this "truth" is limited, and we shall consider the extent of its limitation.

---

[1] It follows that a natural object is associated also with a straight line. Three points A, B and C on a rigid body thus lie in a straight line when, the points A and C being given, B is chosen such that the sum of the distances AB and BC is as short as possible. This incomplete suggestion will suffice for our present purpose.

2 THE SYSTEM OF COORDINATES

On the basis of the physical interpretation of distance which has been indicated, we are also in a position to establish the distance between two points on a rigid body by means of measurements. For this purpose we require a "distance" (rod S) which is to be used once and for all, and which we employ as a standard measure. If, now, A and B are two points on a rigid body, we can construct the line joining them according to the rules of geometry; then, starting from A, we can mark off the distance S time after time until we reach B. The number of these operations required is the numerical measure of the distance AB. This is the basis of all measurement of length.[1]

Every description of the scene of an event or of the position of an object in space is based on the specification of the point on a rigid body (body of reference) with which that event or object coincides. This applies not only to scientific description, but also to everyday life. If I analyse the place specification "Trafalgar Square, London,"[2] I arrive at the following result. The earth is the rigid body to which the specification of place refers; "Trafalgar Square, London," is a well-defined point, to which a name has been assigned, and with which the event coincides in space.[3]

This primitive method of place specification deals only with places on the surface of rigid bodies, and is dependent on the existence of points on this surface which are distinguishable from each other. But we can free ourselves from both of these limitations without altering the nature of our specification of position. If, for instance, a cloud is hovering over Trafalgar Square, then we can determine its position relative to the surface of the earth by erecting a pole perpendicularly on the Square, so that it reaches the cloud. The length of the pole measured with the standard measuring-rod, combined with the specification of the position of the foot of the pole, supplies us with a complete place specification. On the basis of this illustration, we are able to see the manner in which a refinement of

---

[1]Here we have assumed that there is nothing left over, i.e., that the measurement gives a whole number. This difficulty is got over by the use of divided measuring-rods, the introduction of which does not demand any fundamentally new method.

[2]I have chosen this as being more familiar to the English reader than the "Potsdamer Platz, Berlin," which is referred to in the original. (R. W. L.)

[3]It is not necessary here to investigate further the significance of the expression "coincidence in space." This conception is sufficiently obvious to ensure that differences of opinion are scarcely likely to arise as to its applicability in practice.

11

the conception of position has been developed.

(a) We imagine the rigid body, to which the place specification is referred, supplemented in such a manner that the object whose position we require is reached by the completed rigid body.

(b) In locating the position of the object, we make use of a number (here the length of the pole measured with the measuring-rod) instead of designated points of reference.

(c) We speak of the height of the cloud even when the pole which reaches the cloud has not been erected. By means of optical observations of the cloud from different positions on the ground, and taking into account the properties of the propagation of light, we determine the length of the pole we should have required in order to reach the cloud.

From this consideration we see that it will be advantageous if, in the description of position, it should be possible by means of numerical measures to make ourselves independent of the existence of marked positions (possessing names) on the rigid body of reference. In the physics of measurement this is attained by the application of the Cartesian system of coordinates.

This consists of three plane surfaces perpendicular to each other and rigidly attached to a rigid body. Referred to a system of coordinates, the scene of any event will be determined (for the main part) by the specification of the lengths of the three perpendiculars or coordinates (x, y, z) which can be dropped from the scene of the event to those three plane surfaces. The lengths of these three perpendiculars can be determined by a series of manipulations with rigid measuring-rods performed according to the rules and methods laid down by Euclidean geometry.

In practice, the rigid surfaces which constitute the system of coordinates are generally not available; furthermore, the magnitudes of the coordinates are not actually determined by constructions with rigid rods, but by indirect means. If the results of physics and astronomy are to maintain their clearness, the physical meaning of specifications of position must always be sought in accordance with the above considerations.[4]

We thus obtain the following result: Every description of events in space involves the use of a rigid body to which such events have to be referred. The resulting relationship takes for granted that the laws of Euclidean geometry hold for "distances," the "distance" being represented physically by means of the convention of two marks on a rigid body.

---

[4] A refinement and modification of these views does not become necessary until we come to deal with the general theory of relativity, treated in the second part of this book.

3 SPACE AND TIME IN CLASSICAL MECHANICS

The purpose of mechanics is to describe how bodies change their position in space with "time." I should load my conscience with grave sins against the sacred spirit of lucidity were I to formulate the aims of mechanics in this way, without serious reflection and detailed explanations. Let us proceed to disclose these sins.

It is not clear what is to be understood here by "position" and "space." I stand at the window of a railway carriage which is travelling uniformly, and drop a stone on the embankment, without throwing it. Then, disregarding the influence of the air resistance, I see the stone descend in a straight line. A pedestrian who observes the misdeed from the footpath notices that the stone falls to earth in a parabolic curve. I now ask: Do the "positions" traversed by the stone lie "in reality" on a straight line or on a parabola? Moreover, what is meant here by motion "in space"? From the considerations of the previous section the answer is self-evident. In the first place, we entirely shun the vague word "space," of which, we must honestly acknowledge, we cannot form the slightest conception, and we replace it by "motion relative to a practically rigid body of reference." The positions relative to the body of reference (railway carriage or embankment) have already been defined in detail in the preceding section. If instead of "body of reference" we insert "system of coordinates," which is a useful idea for mathematical description, we are in a position to say: The stone traverses a straight line relative to a system of coordinates rigidly attached to the carriage, but relative to a system of coordinates rigidly attached to the ground (embankment) it describes a parabola. With the aid of this example it is clearly seen that there is no such thing as an independently existing trajectory (lit. "path-curve"[1]), but only a trajectory relative to a particular body of reference.

In order to have a *complete* description of the motion, we must specify how the body alters its position *with time*; i.e., for every point on the trajectory it must be stated at what time the body is situated there. These data must be supplemented by such a definition of time that, in virtue of this definition, these time-values can be regarded essentially as magnitudes (results of measurements) capable of observation. If we take our stand on the ground of classical mechanics, we can satisfy this requirement for our illustration in the following manner. We imagine two clocks of identical con-

---

[1]That is, a curve along which the body moves.

struction; the man at the railway-carriage window is holding one of them, and the man on the footpath the other. Each of the observers determines the position on his own reference-body occupied by the stone at each tick of the clock he is holding in his hand. In this connection we have not taken account of the inaccuracy involved by the finiteness of the velocity of propagation of light. With this and with a second difficulty prevailing here we shall have to deal in detail later.

4 THE GALILEIAN SYSTEM OF COORDINATES

As is well known, the fundamental law of the mechanics of Galilei-Newton, which is known as the *law of inertia*, can be stated thus: A body removed sufficiently far from other bodies continues in a state of rest or of uniform motion in a straight line. This law not only says something about the motion of the bodies, but it also indicates the reference-bodies or systems of coordinates, permissible in mechanics, which can be used in mechanical description. The visible fixed stars are bodies for which the law of inertia certainly holds to a high degree of approximation. Now if we use a system of coordinates which is rigidly attached to the earth, then, relative to this system, every fixed star describes a circle of immense radius in the course of an astronomical day, a result which is opposed to the statement of the law of inertia. So that if we adhere to this law we must refer these motions only to systems of coordinates relative to which the fixed stars do not move in a circle. A system of coordinates of which the state of motion is such that the law of inertia holds relative to it is called a "Galileian system of coordinates." The laws of the mechanics of Galilei-Newton can be regarded as valid only for a Galileian system of coordinates.

5 THE PRINCIPLE OF RELATIVITY (IN THE RESTRICTED SENSE)

In order to attain the greatest possible clearness, let us return to our example of the railway carriage supposed to be travelling uniformly. We call its motion a uniform translation ("uniform" because it is of constant velocity and direction, "translation" because although the carriage changes its position relative to the embankment yet it does not rotate in so doing). Let us imagine a raven flying through the air in such a manner that its motion, as observed from the embankment, is uniform and in a straight line. If we were to observe the flying raven from the moving railway carriage, we should find that the motion of the raven would be one of different velocity and direction, but that it would still be uniform and in a straight line. Expressed in an abstract manner we may say: If a mass m is moving uniformly in a straight line with respect to a coordinate system K, then it will also be moving uniformly and in a straight line relative to a second coordinate system K', provided that the latter is executing a uniform translatory motion with respect to K. In accordance with the discussion contained in the preceding chapter, it follows that:

If K is a Galileian coordinate system, then every other coordinate system K' is a Galileian one, when, in relation to K, it is in a condition of uniform motion of translation. Relative to K' the mechanical laws of Galilei-Newton hold good exactly as they do with respect to K.

We advance a step farther in our generalisation when we express the tenet thus: If, relative to K, K' is a uniformly moving coordinate system devoid of rotation, then natural phenomena run their course with respect to K' according to exactly the same general laws as with respect to K. This statement is called the *principle of relativity* (in the restricted sense).

As long as one was convinced that all natural phenomena were capable of representation with the help of classical mechanics, there was no need to doubt the validity of this principle of relativity. But in view of the more recent development of electrodynamics and optics it became more and more evident that classical mechanics affords an insufficient foundation for the physical description of all natural phenomena. At this juncture the question of the validity of the principle of relativity became ripe for discussion, and it did not appear impossible that the answer to this question might be in the negative.

Nevertheless, there are two general facts which at the outset speak very much in favour of the validity of the principle of relativity. Even though classical mechanics does not supply us with a sufficiently broad basis for the theoretical presentation of all physical phenomena, still we must grant it a considerable measure of "truth," since it supplies us with the actual motions of the heavenly bodies with a delicacy of detail little short of wonderful. The principle of relativity must therefore apply with great accuracy in the domain of *mechanics*. But that a principle of such broad generality should hold with such exactness in one domain of phenomena, and yet should be invalid for another, is *a priori* not very probable.

We now proceed to the second argument, to which, moreover, we shall return later. If the principle of relativity (in the restricted sense) does not hold, then the Galileian coordinate systems K, K', K'', etc., which are moving uniformly relative to each other, will not be *equivalent* for the description of natural phenomena. In this case we should be constrained to believe that natural laws are capable of being formulated in a particularly simple manner, and of course only on condition that, from amongst all possible Galileian coordinate systems, we should have chosen *one* (K_0) of a particular state of motion as our body of reference. We should then be justified (because of its merits for the description of natural phenomena) in calling this system "absolutely at rest," and all other Galileian systems K "in motion." If, for instance, our embankment were the system K_0, then our railway carriage would be a system K, relative to which less simple laws would hold than with respect to K_0. This diminished simplicity would be due to the fact that the carriage K would be in motion (i.e., "really") with respect to K_0. In the general laws of nature which have been formulated with reference to K, the magnitude and direction of the velocity of the carriage would necessarily play a part. We should expect, for instance, that the note emitted by an organ-pipe placed with its axis parallel to the direction of travel would be different from that emitted if the axis of the pipe were placed perpendicular to this direction. Now in virtue of its motion in an orbit round the sun, our earth is comparable with a railway carriage travelling with a velocity of about 30 kilometres per second. If the principle of relativity were not valid we should therefore expect that the direction of motion of the earth at any moment would enter into the laws of nature, and also that physical systems in their behaviour would be dependent on the orientation in space with respect to the earth. For owing to the alteration in direction of the velocity of revolution of the earth in the course of a year, the earth cannot be at rest relative to the hypothetical system K_0 throughout the whole year. However, the most careful observations have never revealed such anisotropic properties in terrestrial physical space, i.e., a physical non-equivalence of different directions. This is a very powerful argument in favour of the principle of relativity.

6 THE THEOREM OF THE ADDITION OF VELOCITIES EMPLOYED IN CLASSICAL MECHANICS

Let us suppose our old friend the railway carriage to be travelling along the rails with a constant velocity v, and that a man traverses the length of the carriage in the direction of travel with a velocity w. How quickly or, in other words, with what velocity W does the man advance relative to the embankment during the process? The only possible answer seems to result from the following consideration: If the man were to stand still for a second, he would advance relative to the embankment through a distance v equal numerically to the velocity of the carriage. As a consequence of his walking, however, he traverses an additional distance w relative to the carriage, and hence also relative to the embankment, in this second, the distance w being numerically equal to the velocity with which he is walking. Thus in total he covers the distance $W = v + w$ relative to the embankment in the second considered. We shall see later that this result, which expresses the theorem of the addition of velocities employed in classical mechanics, cannot be maintained; in other words, the law that we have just written down does not hold in reality. For the time being, however, we shall assume its correctness.

7 THE APPARENT INCOMPATIBILITY OF THE LAW OF PROPAGATION OF LIGHT WITH THE PRINCIPLE OF RELATIVITY

There is hardly a simpler law in physics than that according to which light is propagated in empty space. Every child at school knows, or believes he knows, that this propagation takes place in straight lines with a velocity $c = 300,000$ km./sec. At all events we know with great exactness that this velocity is the same for all colours, because if this were not the case, the minimum of emission would not be observed simultaneously for different colours during the eclipse of a fixed star by its dark neighbour. By means of similar considerations based on observations of double stars, the Dutch astronomer De Sitter was also able to show that the velocity of propagation of light cannot depend on the velocity of motion of the body emitting the light. The assumption that this velocity of propagation is dependent on the direction "in space" is in itself improbable.

In short, let us assume that the simple law of the constancy of the velocity of light c (in vacuum) is justifiably believed by the child at school. Who would imagine that this simple law has plunged the conscientiously thoughtful physicist into the greatest intellectual difficulties? Let us consider how these difficulties arise.

Of course we must refer the process of the propagation of light (and indeed every other process) to a rigid reference-body (coordinate system). As such a system let us again choose our embankment. We shall imagine the air above it to have been removed. If a ray of light be sent along the embankment, we see from the above that the tip of the ray will be transmitted with the velocity c relative to the embankment. Now let us suppose that our railway carriage is again travelling along the railway lines with the velocity v, and that its direction is the same as that of the ray of light, but its velocity of course much less. Let us inquire about the velocity of propagation of the ray of light relative to the carriage. It is obvious that we can here apply the consideration of the previous chapter, since the ray of light plays the part of the man walking along relatively to the carriage. The velocity W of the man relative to the embankment is here replaced by the velocity of light relative to the embankment. w is the required velocity of light with respect to the carriage, and we have

$$w = c - v.$$

19

The velocity of propagation of a ray of light relative to the carriage thus comes out smaller than c.

But this result comes into conflict with the principle of relativity set forth in Chapter 5. For, like every other general law of nature, the law of the transmission of light *in vacuo* must, according to the principle of relativity, be the same for the railway carriage as reference-body as when the rails are the body of reference. But, from our above consideration, this would appear to be impossible. If every ray of light is propagated relative to the embankment with the velocity c, then for this reason it would appear that another law of propagation of light must necessarily hold with respect to the carriage—a result contradictory to the principle of relativity.

In view of this dilemma there appears to be nothing else for it than to abandon either the principle of relativity or the simple law of the propagation of light *in vacuo*. Those of you who have carefully followed the preceding discussion are almost sure to expect that we should retain the principle of relativity, which appeals so convincingly to the intellect because it is so natural and simple. The law of the propagation of light *in vacuo* would then have to be replaced by a more complicated law conformable to the principle of relativity. The development of theoretical physics shows, however, that we cannot pursue this course. The epoch-making theoretical investigations of H. A. Lorentz on the electrodynamical and optical phenomena connected with moving bodies show that experience in this domain leads conclusively to a theory of electromagnetic phenomena, of which the law of the constancy of the velocity of light *in vacuo* is a necessary consequence. Prominent theoretical physicists were therefore more inclined to reject the principle of relativity, in spite of the fact that no empirical data had been found which were contradictory to this principle.

At this juncture the theory of relativity entered the arena. As a result of an analysis of the physical conceptions of time and space, it became evident that *in reality there is not the least incompatibility between the principle of relativity and the law of propagation of light,* and that by systematically holding fast to both these laws a logically rigid theory could be arrived at. This theory has been called the *special theory of relativity* to distinguish it from the extended theory, with which we shall deal later. In the following pages we shall present the fundamental ideas of the special theory of relativity.

8 ON THE IDEA OF TIME IN PHYSICS

Lightning has struck the rails on our railway embankment at two places A and B far distant from each other. I make the additional assertion that these two lightning flashes occurred simultaneously. If I ask you whether there is sense in this statement, you will answer my question with a decided "Yes." But if I now approach you with the request to explain to me the sense of the statement more precisely, you find after some consideration that the answer to this question is not so easy as it appears at first sight.

After some time perhaps the following answer would occur to you: "The significance of the statement is clear in itself and needs no further explanation; of course it would require some consideration if I were to be commissioned to determine by observations whether in the actual case the two events took place simultaneously or not." I cannot be satisfied with this answer for the following reason. Supposing that as a result of ingenious considerations an able meteorologist were to discover that the lightning must always strike the places A and B simultaneously, then we should be faced with the task of testing whether or not this theoretical result is in accordance with the reality. We encounter the same difficulty with all physical statements in which the conception "simultaneous" plays a part. The concept does not exist for the physicist until he has the possibility of discovering whether or not it is fulfilled in an actual case. We thus require a definition of simultaneity such that this definition supplies us with the method by means of which, in the present case, he can decide by experiment whether or not both the lightning strokes occurred simultaneously. As long as this requirement is not satisfied, I allow myself to be deceived as a physicist (and of course the same applies if I am not a physicist), when I imagine that I am able to attach a meaning to the statement of simultaneity. (I would ask the reader not to proceed farther until he is fully convinced on this point.)

After thinking the matter over for some time you then offer the following suggestion with which to test simultaneity. By measuring along the rails, the connecting line AB should be measured up and an observer placed at the mid-point M of the distance AB. This observer should be supplied with an arrangement (e.g., two mirrors inclined at 90^0) which allows him visually to observe both places A and B at the same time. If the observer perceives the two flashes of lightning at the same time, then they are simultaneous.

I am very pleased with this suggestion, but for all that I cannot regard the matter as quite settled, because I feel constrained to raise the following objection: "Your definition would certainly be right, if I only knew that the light by means of which the observer at M perceives the lightning flashes travels along the length $A \longrightarrow M$ with the same velocity as along the length $B \longrightarrow M$. But an examination of this supposition would only be possible if we already had at our disposal the means of measuring time. It would thus appear as though we were moving here in a logical circle."

After further consideration you cast a somewhat disdainful glance at me—and rightly so—and you declare: "I maintain my previous definition nevertheless, because in reality it assumes absolutely nothing about light. There is only *one* demand to be made of the definition of simultaneity, namely, that in every real case it must supply us with an empirical decision as to whether or not the conception that has to be defined is fulfilled. That my definition satisfies this demand is indisputable. That light requires the same time to traverse the path $A \longrightarrow M$ as for the path $B \longrightarrow M$ is in reality neither a *supposition nor a hypothesis* about the physical nature of light, but a *stipulation* which I can make of my own freewill in order to arrive at a definition of simultaneity."

It is clear that this definition can be used to give an exact meaning not only to *two* events, but to as many events as we care to choose, and independently of the positions of the scenes of the events with respect to the body of reference[1] (here the railway embankment). We are thus led also to a definition of "time" in physics. For this purpose we suppose that clocks of identical construction are placed at the points A, B and C of the railway line (coordinate system), and that they are set in such a manner that the positions of their pointers are simultaneously (in the above sense) the same. Under these conditions we understand by the "time" of an event the reading (position of the hands) of that one of these clocks which is in the immediate vicinity (in space) of the event. In this manner a time-value is associated with every event which is essentially capable of observation.

This stipulation contains a further physical hypothesis, the validity of which will hardly be doubted without empirical evidence to the contrary. It has been assumed that all these clocks go *at the same rate* if they are of identical construction. Stated more exactly: When two clocks arranged at rest in different places of a reference-body are set in such a manner that a *particular* position of the pointers of the one clock is *simultaneous* (in the above sense) with the *same* position of the pointers of the other clock, then identical "settings" are always simultaneous (in the sense of the above definition).

---

[1] We suppose further, that, when three events A, B and C occur in different places in such a manner that A is simultaneous with B, and B is simultaneous with C (simultaneous in the sense of the above definition), then the criterion for the simultaneity of the pair of events A, C is also satisfied. This assumption is a physical hypothesis about the law of propagation of light; it must certainly be fulfilled if we are to maintain the law of the constancy of the velocity of light *in vacuo*.

9 THE RELATIVITY OF SIMULTANEITY

Up to now our considerations have been referred to a particular body of reference, which we have styled a "railway embankment." We suppose a very long train travelling along the rails with the constant velocity v and in the direction indicated in Fig. 1. People travelling in this train will with advantage use the train as a rigid reference-body (coordinate system); they regard all events in reference to the train. Then every event which takes place along the line also takes place at a particular point of the train. Also the definition of simultaneity can be given relative to the train in exactly the same way as with respect to the embankment. As a natural consequence, however, the following question arises:

Are two events (e.g., the two strokes of lightning A and B) which are simultaneous *with reference to the railway embankment* also simultaneous *relatively to the train*? We shall show directly that the answer must be in the negative.

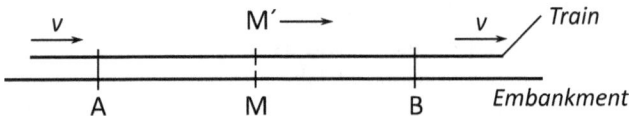

Fig. 1

When we say that the lightning strokes A and B are simultaneous with respect to the embankment, we mean: the rays of light emitted at the places A and B, where the lightning occurs, meet each other at the mid-point M of the length $A \longrightarrow B$ of the embankment. But the events A and B also correspond to positions A and B on the train. Let M' be the mid-point of the distance $A \longrightarrow B$ on the travelling train. Just when the flashes[1] of lightning occur, this point M' naturally coincides with the point M, but it moves towards the right in the diagram with the velocity v of the train. If an observer sitting in the position M' in the train did not possess this velocity, then he would remain permanently at M, and the light rays emitted by the flashes of lightning A and B would reach him simultaneously, i.e., they would meet just where he is situated. Now in reality (considered with reference to the railway embankment) he is hastening towards the beam of light coming from B, whilst he is riding on ahead of the beam of

---

[1] As judged from the embankment.

light coming from A. Hence the observer will see the beam of light emitted from B earlier than he will see that emitted from A. Observers who take the railway train as their reference-body must therefore come to the conclusion that the lightning flash B took place earlier than the lightning flash A. We thus arrive at the important result:

Events which are simultaneous with reference to the embankment are not simultaneous with respect to the train, and *vice versa* (relativity of simultaneity). Every reference-body (coordinate system) has its own particular time; unless we are told the reference-body to which the statement of time refers, there is no meaning in a statement of the time of an event.

Now before the advent of the theory of relativity it had always tacitly been assumed in physics that the statement of time had an absolute significance, i.e., that it is independent of the state of motion of the body of reference. But we have just seen that this assumption is incompatible with the most natural definition of simultaneity; if we discard this assumption, then the conflict between the law of the propagation of light *in vacuo* and the principle of relativity (developed in Chapter 7) disappears.

We were led to that conflict by the considerations of Chapter 6, which are now no longer tenable. In that chapter we concluded that the man in the carriage, who traverses the distance w *per second* relative to the carriage, traverses the same distance also with respect to the embankment *in each second* of time. But, according to the foregoing considerations, the time required by a particular occurrence with respect to the carriage must not be considered equal to the duration of the same occurrence as judged from the embankment (as reference-body). Hence it cannot be contended that the man in walking travels the distance w relative to the railway line in a time which is equal to one second as judged from the embankment.

Moreover, the considerations of Chapter 6 are based on yet a second assumption, which, in the light of a strict consideration, appears to be arbitrary, although it was always tacitly made even before the introduction of the theory of relativity.

10 ON THE RELATIVITY OF THE CONCEPTION OF DISTANCE

Let us consider two particular points on the train[1] travelling along the embankment with the velocity v, and inquire as to their distance apart. We already know that it is necessary to have a body of reference for the measurement of a distance, with respect to which body the distance can be measured up. It is the simplest plan to use the train itself as reference-body (coordinate system). An observer in the train measures the interval by marking off his measuring-rod in a straight line (e.g., along the floor of the carriage) as many times as is necessary to take him from the one marked point to the other. Then the number which tells us how often the rod has to be laid down is the required distance.

It is a different matter when the distance has to be judged from the railway line. Here the following method suggests itself. If we call A' and B' the two points on the train whose distance apart is required, then both of these points are moving with the velocity v along the embankment. In the first place we require to determine the points A and B of the embankment which are just being passed by the two points A' and B' at a particular time t—judged from the embankment. These points A and B of the embankment can be determined by applying the definition of time given in Chapter 8. The distance between these points A and B is then measured by repeated application of the measuring-rod along the embankment.

A priori it is by no means certain that this last measurement will supply us with the same result as the first. Thus the length of the train as measured from the embankment may be different from that obtained by measuring in the train itself. This circumstance leads us to a second objection which must be raised against the apparently obvious consideration of Chapter 6. Namely, if the man in the carriage covers the distance w in a unit of time—*measured from the train*,—then this distance—*as measured from the embankment*—is not necessarily also equal to w.

---

[1] E.g. the middle of the first and of the hundredth carriage.

11 THE LORENTZ TRANSFORMATION

The results of the last three chapters show that the apparent incompatibility of the law of propagation of light with the principle of relativity (Chapter 7) has been derived by means of a consideration which borrowed two unjustifiable hypotheses from classical mechanics; these are as follows:

(1) The time-interval (time) between two events is independent of the condition of motion of the body of reference.

(2) The space-interval (distance) between two points of a rigid body is independent of the condition of motion of the body of reference.

If we drop these hypotheses, then the dilemma of Chapter 7 disappears, because the theorem of the addition of velocities derived in Chapter 6 becomes invalid. The possibility presents itself that the law of the propagation of light *in vacuo* may be compatible with the principle of relativity, and the question arises: How have we to modify the considerations of Chapter 6 in order to remove the apparent disagreement between these two fundamental results of experience? This question leads to a general one. In the discussion of Chapter 6 we have to do with places and times relative both to the train and to the embankment. How are we to find the place and time of an event in relation to the train, when we know the place and time of the event with respect to the railway embankment? Is there a thinkable answer to this question of such a nature that the law of transmission of light *in vacuo* does not contradict the principle of relativity? In other words: Can we conceive of a relation between place and time of the individual events relative to both reference-bodies, such that every ray of light possesses the velocity of transmission c relative to the embankment and relative to the train? This question leads to a quite definite positive answer, and to a perfectly definite transformation law for the space-time magnitudes of an event when changing over from one body of reference to another.

Before we deal with this, we shall introduce the following incidental consideration. Up to the present we have only considered events taking place along the embankment, which had mathematically to assume the function of a straight line. In the manner indicated in Chapter 2 we can imagine this reference-body supplemented laterally and in a vertical direction by means of a framework of rods, so that an event which takes place anywhere can be localised with reference to this framework. Similarly, we can imagine the train travelling with the velocity v to be continued across the whole of space, so that every event, no matter how far off it may be, could also be localised with respect to the second framework. Without committing any fundamental error, we can disregard the fact that in reality these frameworks would continually interfere with each other, owing to the impenetrability of solid bodies. In every such framework we imagine three surfaces perpendicular to each other marked out, and designated as "coordinate planes" ("coordinate system"). A coordinate system K then corresponds to the embankment, and a coordinate system K' to the train. An event, wherever it may have taken place, would be fixed in space with respect to K by the three perpendiculars x, y, z on the coordinate planes, and with regard to time by a time-value t. Relative to K', *the same event* would be fixed in respect of space and time by corresponding values x', y', z', t', which of course are not identical with x, y, z, t. It has already been set forth in detail how these magnitudes are to be regarded as results of physical measurements.

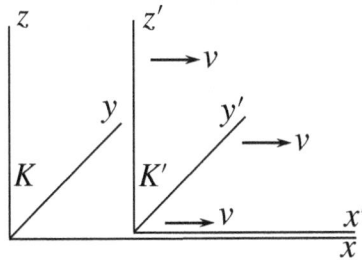

Fig. 2

Obviously our problem can be exactly formulated in the following manner. What are the values x', y', z', t', of an event with respect to K', when the magnitudes x, y, z, t, of the same event with respect to K are given? The relations must be so chosen that the law of the transmission of light *in vacuo* is satisfied for one and the same ray of light (and of course for every ray) with respect to K and K'. For the relative orientation in space of the coordinate systems indicated in the diagram (Fig. 2), this problem is solved by means of the equations:

$$x' = \frac{x - vt}{\sqrt{1 - \dfrac{v^2}{c^2}}}$$

$$y' = y$$
$$z' = z$$

$$t' = \frac{t - \dfrac{v}{c^2}x}{\sqrt{1 - \dfrac{v^2}{c^2}}}.$$

This system of equations is known as the "Lorentz transformation."[1]

If in place of the law of transmission of light we had taken as our basis the tacit assumptions of the older mechanics as to the absolute character of times and lengths, then instead of the above we should have obtained the following equations:

$$x' = x - vt$$
$$y' = y$$
$$z' = z$$
$$t' = t.$$

This system of equations is often termed the "Galilei transformation." The Galilei transformation can be obtained from the Lorentz transformation by substituting an infinitely large value for the velocity of light c in the latter transformation.

Aided by the following illustration, we can readily see that, in accordance with the Lorentz transformation, the law of the transmission of light *in vacuo* is satisfied both for the reference-body K and for the reference-body K'. A light-signal is sent along the positive x-axis, and this light-stimulus advances in accordance with the equation

$$x = ct,$$

i.e., with the velocity c. According to the equations of the Lorentz transformation, this simple relation between x and t involves a relation between x' and t'. In point of fact, if we substitute for x the value ct in the first and fourth equations of the Lorentz transformation, we obtain:

$$x' = \frac{(c - v)t}{\sqrt{1 - \frac{v^2}{c^2}}}$$

$$t' = \frac{\left(1 - \frac{v}{c}\right)t}{\sqrt{1 - \frac{v^2}{c^2}}},$$

from which, by division, the expression

$$x' = ct'$$

immediately follows. If referred to the system K', the propagation of light takes place according to this equation. We thus see that the velocity of transmission relative to the reference-body K' is also equal to c. The same result is obtained for rays of light advancing in any other direction whatsoever. Of course this is not surprising, since the equations of the Lorentz transformation were derived conformably to this point of view.

---

[1] A simple derivation of the Lorentz transformation is given in Appendix 1.

12 THE BEHAVIOUR OF MEASURING-RODS AND CLOCKS IN MOTION

I place a metre-rod in the x'-axis of K' in such a manner that one end (the beginning) coincides with the point $x' = 0$, whilst the other end (the end of the rod) coincides with the point $x' = 1$. What is the length of the metre-rod relatively to the system K? In order to learn this, we need only ask where the beginning of the rod and the end of the rod lie with respect to K at a particular time t of the system K. By means of the first equation of the Lorentz transformation the values of these two points at the time $t = 0$ can be shown to be

$$x_{\text{(beginning of rod)}} = 0 \sqrt{1 - \frac{v^2}{c^2}}$$

$$x_{\text{(end of rod)}} = 1 \sqrt{1 - \frac{v^2}{c^2}},$$

the distance between the points being $\sqrt{1 - v^2/c^2}$. But the metre-rod is moving with the velocity v relative to K. It therefore follows that the length of a rigid metre-rod moving in the direction of its length with a velocity v is $\sqrt{1 - v^2/c^2}$ of a metre. The rigid rod is thus shorter when in motion than when at rest, and the more quickly it is moving, the shorter is the rod. For the velocity $v = c$ we should have $\sqrt{1 - v^2/c^2} = 0$, and for still greater velocities the square-root becomes imaginary. From this we conclude that in the theory of relativity the velocity c plays the part of a limiting velocity, which can neither be reached nor exceeded by any real body.

Of course this feature of the velocity c as a limiting velocity also clearly follows from the equations of the Lorentz transformation, for these become meaningless if we choose values of v greater than c.

If, on the contrary, we had considered a metre-rod at rest in the x-axis with respect to K, then we should have found that the length of the rod as judged from K' would have been $\sqrt{1 - v^2/c^2}$; this is quite in accordance with the principle of relativity which forms the basis of our considerations.

A priori it is quite clear that we must be able to learn something about the physical behaviour of measuring-rods and clocks from the equations of transformation, for the magnitudes x, y, z, t, are nothing more nor less than the results of measurements obtainable by means of measuring-rods and clocks. If we had based our considerations on the Galilei transformation we should not have obtained a contraction of the rod as a consequence of its motion.

Let us now consider a seconds-clock which is permanently situated at the origin ($x' = 0$) of K'. $t' = 0$ and $t' = 1$ are two successive ticks of this clock. The first and fourth equations of the Lorentz transformation give for these two ticks:

$$t = 0$$

and

$$t = \frac{1}{\sqrt{1 - \dfrac{v^2}{c^2}}}.$$

As judged from K, the clock is moving with the velocity v; as judged from this reference-body, the time which elapses between two strokes of the clock is not one second, but $\dfrac{1}{\sqrt{1 - v^2/c^2}}$ seconds, i.e., a somewhat larger time. As a consequence of its motion the clock goes more slowly than when at rest. Here also the velocity c plays the part of an unattainable limiting velocity.

13 Theorem of the Addition of Velocities. The Experiment of Fizeau

Now in practice we can move clocks and measuring-rods only with velocities that are small compared with the velocity of light; hence we shall hardly be able to compare the results of the previous chapter directly with the reality. But, on the other hand, these results must strike you as being very singular, and for that reason I shall now draw another conclusion from the theory, one which can easily be derived from the foregoing considerations, and which has been most elegantly confirmed by experiment.

In Chapter 6 we derived the theorem of the addition of velocities in one direction in the form which also results from the hypotheses of classical mechanics. This theorem can also be deduced readily from the Galilei transformation (Chapter 11). In place of the man walking inside the carriage, we introduce a point moving relatively to the coordinate system K' in accordance with the equation

$$x' = wt'.$$

By means of the first and fourth equations of the Galilei transformation we can express x' and t' in terms of x and t, and we then obtain

$$x = (v + w)t.$$

This equation expresses nothing else than the law of motion of the point with reference to the system K (of the man with reference to the embankment). We denote this velocity by the symbol W, and we then obtain, as in Chapter 6,

$$W = v + w. \tag{A}$$

But we can carry out this consideration just as well on the basis of the theory of relativity. In the equation

$$x' = wt'$$

we must then express x' and t' in terms of x and t, making use of the first and fourth equations of the *Lorentz transformation*. Instead of the equation (A) we then obtain the equation

$$W = \frac{v + w}{1 + \dfrac{vw}{c^2}}, \tag{B}$$

31

which corresponds to the theorem of addition for velocities in one direction according to the theory of relativity. The question now arises as to which of these two theorems is the better in accord with experience. On this point we are enlightened by a most important experiment which the brilliant physicist Fizeau performed more than half a century ago, and which has been repeated since then by some of the best experimental physicists, so that there can be no doubt about its result. The experiment is concerned with the following question. Light travels in a motionless liquid with a particular velocity w. How quickly does it travel in the direction of the arrow in the tube T (see the accompanying diagram, Fig. 3) when the liquid above mentioned is flowing through the tube with a velocity v?

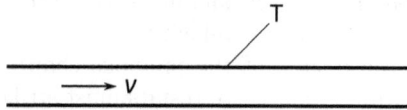

Fig. 3

In accordance with the principle of relativity we shall certainly have to take for granted that the propagation of light always takes place with the same velocity w *with respect to the liquid*, whether the latter is in motion with reference to other bodies or not. The velocity of light relative to the liquid and the velocity of the latter relative to the tube are thus known, and we require the velocity of light relative to the tube.

It is clear that we have the problem of Chapter 6 again before us. The tube plays the part of the railway embankment or of the coordinate system K, the liquid plays the part of the carriage or of the coordinate system K', and finally, the light plays the part of the man walking along the carriage, or of the moving point in the present chapter. If we denote the velocity of the light relative to the tube by W, then this is given by the equation (A) or (B), according as the Galilei transformation or the Lorentz transformation corresponds to the facts. Experiment[1] decides in favour of equation (B) derived from the theory of relativity, and the agreement is, indeed, very exact. According to recent and most excellent measurements by Zeeman, the influence of the velocity of flow v on the propagation of light is represented by formula (B) to within one per cent.

Nevertheless we must now draw attention to the fact that a theory of this phenomenon was given by H. A. Lorentz long before the statement of the theory of relativity. This theory was of a purely electrodynamical nature, and was obtained by the use of particular hypotheses as to the electromagnetic structure of matter. This circumstance, however, does not in the least diminish the conclusiveness of the experiment as a crucial test in favour of the theory of relativity, for the electrodynamics of Maxwell-

---

[1] Fizeau found $W = w + v\left(1 - 1/n^2\right)$, where $n = c/w$ is the index of refraction of the liquid. On the other hand, owing to the smallness of vw/c^2 as compared with 1, we can replace (B) in the first place by $W = (w + v)\left(1 - vw/c^2\right)$, or to the same order of approximation by $w + v\left(1 - 1/n^2\right)$, which agrees with Fizeau's result.

Lorentz, on which the original theory was based, in no way opposes the theory of relativity. Rather has the latter been developed from electrodynamics as an astoundingly simple combination and generalisation of the hypotheses, formerly independent of each other, on which electrodynamics was built.

14 The Heuristic Value of the Theory of Relativity

Our train of thought in the foregoing pages can be epitomised in the following manner. Experience has led to the conviction that, on the one hand, the principle of relativity holds true, and that on the other hand the velocity of transmission of light *in vacuo* has to be considered equal to a constant c. By uniting these two postulates we obtained the law of transformation for the rectangular coordinates x, y, z and the time t of the events which constitute the processes of nature. In this connection we did not obtain the Galilei transformation, but, differing from classical mechanics, the *Lorentz transformation*.

The law of transmission of light, the acceptance of which is justified by our actual knowledge, played an important part in this process of thought. Once in possession of the Lorentz transformation, however, we can combine this with the principle of relativity, and sum up the theory thus:

Every general law of nature must be so constituted that it is transformed into a law of exactly the same form when, instead of the space-time variables x, y, z, t of the original coordinate system K, we introduce new space-time variables x', y', z', t' of a coordinate system K'.

In this connection the relation between the ordinary and the accented magnitudes is given by the Lorentz transformation. Or, in brief: General laws of nature are covariant with respect to Lorentz transformations.

This is a definite mathematical condition that the theory of relativity demands of a natural law, and in virtue of this, the theory becomes a valuable heuristic aid in the search for general laws of nature. If a general law of nature were to be found which did not satisfy this condition, then at least one of the two fundamental assumptions of the theory would have been disproved. Let us now examine what general results the latter theory has hitherto evinced.

15 GENERAL RESULTS OF THE THEORY

It is clear from our previous considerations that the (special) theory of relativity has grown out of electrodynamics and optics. In these fields it has not appreciably altered the predictions of theory, but it has considerably simplified the theoretical structure, i.e., the derivation of laws, and—what is incomparably more important—it has considerably reduced the number of independent hypotheses forming the basis of theory. The special theory of relativity has rendered the Maxwell-Lorentz theory so plausible, that the latter would have been generally accepted by physicists even if experiment had decided less unequivocally in its favour.

Classical mechanics required to be modified before it could come into line with the demands of the special theory of relativity. For the main part, however, this modification affects only the laws for rapid motions, in which the velocities of matter v are not very small as compared with the velocity of light. We have experience of such rapid motions only in the case of electrons and ions; for other motions the variations from the laws of classical mechanics are too small to make themselves evident in practice. We shall not consider the motion of stars until we come to speak of the general theory of relativity. In accordance with the theory of relativity the kinetic energy of a material point of mass m is no longer given by the well-known expression

$$m\frac{v^2}{2},$$

but by the expression

$$\frac{mc^2}{\sqrt{1 - \frac{v^2}{c^2}}}.$$

This expression approaches infinity as the velocity v approaches the velocity of light c. The velocity must therefore always remain less than c, however great may be the energies used to produce the acceleration. If we develop the expression for the kinetic energy in the form of a series, we obtain

$$mc^2 + m\frac{v^2}{2} + \frac{3}{8}m\frac{v^4}{c^2} + \ldots.$$

When v^2/c^2 is small compared with unity, the third of these terms is always small in comparison with the second, which last is alone considered in classical mechanics. The first term mc^2 does not contain the velocity, and requires no consideration if we are only dealing with the question as to how the energy of a point-mass depends on the velocity. We shall speak of its essential significance later.

The most important result of a general character to which the special theory of relativity has led is concerned with the conception of mass. Before the advent of relativity, physics recognised two conservation laws of fundamental importance, namely, the law of the conservation of energy and the law of the conservation of mass; these two fundamental laws appeared to be quite independent of each other. By means of the theory of relativity they have been united into one law. We shall now briefly consider how this unification came about, and what meaning is to be attached to it.

The principle of relativity requires that the law of the conservation of energy should hold not only with reference to a coordinate system K, but also with respect to every coordinate system K' which is in a state of uniform motion of translation relative to K, or, briefly, relative to every "Galileian" system of coordinates. In contrast to classical mechanics, the Lorentz transformation is the deciding factor in the transition from one such system to another.

By means of comparatively simple considerations we are led to draw the following conclusion from these premises, in conjunction with the fundamental equations of the electrodynamics of Maxwell: A body moving with the velocity v, which absorbs[1] an amount of energy E_0 in the form of radiation without suffering an alteration in velocity in the process, has, as a consequence, its energy increased by an amount

$$\frac{E_0}{\sqrt{1 - \dfrac{v^2}{c^2}}}.$$

In consideration of the expression given above for the kinetic energy of the body, the required energy of the body comes out to be

$$\frac{\left(m + \dfrac{E_0}{c^2}\right) c^2}{\sqrt{1 - \dfrac{v^2}{c^2}}}.$$

Thus the body has the same energy as a body of mass $\left(m + E_0/c^2\right)$ moving with the velocity v. Hence we can say: If a body takes up an amount of energy E_0, then its inertial mass increases by an amount E_0/c^2; the inertial mass of a body is not a constant, but varies according to the change in the energy of the body.[2] The inertial mass of a system of bodies

---

[1] E_0 is the energy taken up, as judged from a coordinate system moving with the body.

[2] EDITOR'S NOTE: Here Einstein clearly writes "the inertial mass of a body is not a constant" but did not discuss the concept of relativistic mass (i.e., the relativistic in-

can even be regarded as a measure of its energy. The law of the conservation of the mass of a system becomes identical with the law of the conservation of energy, and is only valid provided that the system neither takes up nor sends out energy. Writing the expression for the energy in the form

$$\frac{mc^2 + E_0}{\sqrt{1 - \frac{v^2}{c^2}}},$$

we see that the term mc^2, which has hitherto attracted our attention, is nothing else than the energy possessed by the body[3] before it absorbed the energy E_0.

A direct comparison of this relation with experiment is not possible at the present time, owing to the fact that the changes in energy E_0 to which we can subject a system are not large enough to make themselves perceptible as a change in the inertial mass of the system. E_0/c^2 is too small in comparison with the mass m, which was present before the alteration of the energy. It is owing to this circumstance that classical mechanics was able to establish successfully the conservation of mass as a law of independent validity.

Let me add a final remark of a fundamental nature. The success of the Faraday-Maxwell interpretation of electromagnetic action at a distance resulted in physicists becoming convinced that there are no such things as instantaneous actions at a distance (not involving an intermediary medium) of the type of Newton's law of gravitation. According to the theory of relativity, action at a distance with the velocity of light always takes the place of instantaneous action at a distance or of action at a distance with an infinite velocity of transmission. This is connected with the fact that the velocity c plays a fundamental role in this theory. In Part II we shall see in what way this result becomes modified in the general theory of relativity.

NOTE: With the advent of nuclear transformation processes, which result from the bombardment of elements by α-particles, protons, deuterons, neutrons or γ-rays, the equivalence of mass and energy expressed by the relation $E = mc^2$ has been amply confirmed. The sum of the reacting masses, together with the mass equivalent of the kinetic energy of the bombarding particle (or photon), is always greater than the sum of the resulting masses. The difference is the equivalent mass of the kinetic energy of the

---

crease of the mass of a body as its velocity increases), which he silently abandoned after using it in his 1905 paper where he defined longitudinal mass and transverse mass. As Einstein's unclear view of the relativistic mass might have provided some encouragement for "what has probably been the most vigorous campaign ever waged against the concept of relativistic mass" (M. Jammer, *Concepts of Mass in Contemporary Physics and Philosophy* (Princeton University Press, Princeton 2000) p. 51), the Appendix *On Relativistic Mass* was included in this volume (p. 117) to address that unprecedented campaign and to help the readers make their own decision on the concept of relativistic mass and on that campaign.

[3] As judged from a coordinate system moving with the body.

particles generated, or of the released electromagnetic energy (γ-photons). In the same way, the mass of a spontaneously disintegrating radioactive atom is always greater than the sum of the masses of the resulting atoms by the mass equivalent of the kinetic energy of the particles generated (or of the photonic energy). Measurements of the energy of the rays emitted in nuclear reactions, in combination with the equations of such reactions, render it possible to evaluate atomic weights to a high degree of accuracy.

R.W.L.

16 EXPERIENCE AND THE SPECIAL THEORY OF RELATIVITY

To what extent is the special theory of relativity supported by experience? This question is not easily answered for the reason already mentioned in connection with the fundamental experiment of Fizeau. The special theory of relativity has crystallised out from the Maxwell-Lorentz theory of electromagnetic phenomena. Thus all facts of experience which support the electromagnetic theory also support the theory of relativity. As being of particular importance, I mention here the fact that the theory of relativity enables us to predict the effects produced on the light reaching us from the fixed stars. These results are obtained in an exceedingly simple manner, and the effects indicated, which are due to the relative motion of the earth with reference to those fixed stars, are found to be in accord with experience. We refer to the yearly movement of the apparent position of the fixed stars resulting from the motion of the earth round the sun (aberration), and to the influence of the radial components of the relative motions of the fixed stars with respect to the earth on the colour of the light reaching us from them. The latter effect manifests itself in a slight displacement of the spectral lines of the light transmitted to us from a fixed star, as compared with the position of the same spectral lines when they are produced by a terrestrial source of light (Doppler principle). The experimental arguments in favour of the Maxwell-Lorentz theory, which are at the same time arguments in favour of the theory of relativity, are too numerous to be set forth here. In reality they limit the theoretical possibilities to such an extent, that no other theory than that of Maxwell and Lorentz has been able to hold its own when tested by experience.

But there are two classes of experimental facts hitherto obtained which can be represented in the Maxwell-Lorentz theory only by the introduction of an auxiliary hypothesis, which in itself—i.e., without making use of the theory of relativity—appears extraneous.

It is known that cathode rays and the so-called β-rays emitted by radioactive substances consist of negatively electrified particles (electrons) of very small inertia and large velocity. By examining the deflection of these rays under the influence of electric and magnetic fields, we can study the law of motion of these particles very exactly.

In the theoretical treatment of these electrons, we are faced with the difficulty that electrodynamic theory of itself is unable to give an account

39

of their nature. For since electrical masses of one sign repel each other, the negative electrical masses constituting the electron would necessarily be scattered under the influence of their mutual repulsions, unless there are forces of another kind operating between them, the nature of which has hitherto remained obscure to us.[1] If we now assume that the relative distances between the electrical masses constituting the electron remain unchanged during the motion of the electron (rigid connection in the sense of classical mechanics), we arrive at a law of motion of the electron which does not agree with experience. Guided by purely formal points of view, H. A. Lorentz was the first to introduce the hypothesis that the particles constituting the electron experience a contraction in the direction of motion in consequence of that motion, the amount of this contraction being proportional to the expression $\sqrt{1 - v^2/c^2}$. This hypothesis, which is not justifiable by any electrodynamical facts, supplies us then with that particular law of motion which has been confirmed with great precision in recent years.

The theory of relativity leads to the same law of motion, without requiring any special hypothesis whatsoever as to the structure and the behaviour of the electron. We arrived at a similar conclusion in Chapter 13 in connection with the experiment of Fizeau, the result of which is foretold by the theory of relativity without the necessity of drawing on hypotheses as to the physical nature of the liquid.

The second class of facts to which we have alluded has reference to the question whether or not the motion of the earth in space can be made perceptible in terrestrial experiments. We have already remarked in Chapter 5 that all attempts of this nature led to a negative result. Before the theory of relativity was put forward, it was difficult to become reconciled to this negative result, for reasons now to be discussed. The inherited prejudices about time and space did not allow any doubt to arise as to the prime importance of the Galilei transformation for changing over from one body of reference to another. Now assuming that the Maxwell-Lorentz equations hold for a reference-body K, we then find that they do not hold for a reference-body K' moving uniformly with respect to K, if we assume that the relations of the Galileian transformation exist between the coordinates of K and K'. It thus appears that of all Galileian coordinate systems one (K) corresponding to a particular state of motion is physically unique. This result was interpreted physically by regarding K as at rest with respect to a hypothetical æther of space. On the other hand, all coordinate systems K' moving relatively to K were to be regarded as in motion with respect to the æther. To this motion of K' against the æther ("æther-drift" relative to K') were attributed the more complicated laws which were supposed to hold relative to K'. Strictly speaking, such an æther-drift ought also to be assumed relative to the earth, and for a long time the efforts of physicists were devoted to attempts to detect the existence of an æther-drift

---

[1]The general theory of relativity renders it likely that the electrical masses of an electron are held together by gravitational forces.

at the earth's surface.

In one of the most notable of these attempts Michelson devised a method which appears as though it must be decisive. Imagine two mirrors so arranged on a rigid body that the reflecting surfaces face each other. A ray of light requires a perfectly definite time T to pass from one mirror to the other and back again, if the whole system be at rest with respect to the æther. It is found by calculation, however, that a slightly different time T' is required for this process, if the body, together with the mirrors, be moving relatively to the æther. And yet another point: it is shown by calculation that for a given velocity v with reference to the æther, this time T' is different when the body is moving perpendicularly to the planes of the mirrors from that resulting when the motion is parallel to these planes. Although the estimated difference between these two times is exceedingly small, Michelson and Morley performed an experiment involving interference in which this difference should have been clearly detectable. But the experiment gave a negative result—a fact very perplexing to physicists. Lorentz and FitzGerald rescued the theory from this difficulty by assuming that the motion of the body relative to the æther produces a contraction of the body in the direction of motion, the amount of contraction being just sufficient to compensate for the difference in time mentioned above. Comparison with the discussion in Chapter 12 shows that also from the standpoint of the theory of relativity this solution of the difficulty was the right one. But on the basis of the theory of relativity the method of interpretation is incomparably more satisfactory. According to this theory there is no such thing as a "specially favoured" (unique) coordinate system to occasion the introduction of the æther-idea, and hence there can be no æther-drift, nor any experiment with which to demonstrate it. Here the contraction of moving bodies follows from the two fundamental principles of the theory without the introduction of particular hypotheses; and as the prime factor involved in this contraction we find, not the motion in itself, to which we cannot attach any meaning, but the motion with respect to the body of reference chosen in the particular case in point. Thus for a coordinate system moving with the earth the mirror system of Michelson and Morley is not shortened, but it *is* shortened for a coordinate system which is at rest relatively to the sun.

17   MINKOWSKI'S FOUR-DIMENSIONAL SPACE

The non-mathematician is seized by a mysterious shuddering when he hears of "four-dimensional" things, by a feeling not unlike that awakened by thoughts of the occult. And yet there is no more common-place statement than that the world in which we live is a four-dimensional space-time continuum.

Space is a three-dimensional continuum. By this we mean that it is possible to describe the position of a point (at rest) by means of three numbers (coordinates) x, y, z, and that there is an indefinite number of points in the neighbourhood of this one, the position of which can be described by coordinates such as x_1, y_1, z_1, which may be as near as we choose to the respective values of the coordinates x, y, z of the first point. In virtue of the latter property we speak of a "continuum," and owing to the fact that there are three coordinates we speak of it as being "three-dimensional."

Similarly, the world of physical phenomena which was briefly called "world" by Minkowski is naturally four-dimensional in the space-time sense. For it is composed of individual events, each of which is described by four numbers, namely, three space coordinates x, y, z and a time coordinate, the time-value t. The "world" is in this sense also a continuum; for to every event there are as many "neighbouring" events (realised or at least thinkable) as we care to choose, the coordinates x_1, y_1, z_1, t_1 of which differ by an indefinitely small amount from those of the event x, y, z, t originally considered. That we have not been accustomed to regard the world in this sense as a four-dimensional continuum is due to the fact that in physics, before the advent of the theory of relativity, time played a different and more independent role, as compared with the space coordinates. It is for this reason that we have been in the habit of treating time as an independent continuum. As a matter of fact, according to classical mechanics, time is absolute, i.e., it is independent of the position and the condition of motion of the system of coordinates. We see this expressed in the last equation of the Galileian transformation ($t' = t$).

The four-dimensional mode of consideration of the "world" is natural on the theory of relativity, since according to this theory time is robbed of its independence. This is shown by the fourth equation of the Lorentz transformation:

$$t' = \frac{t - \dfrac{v}{c^2}x}{\sqrt{1 - \dfrac{v^2}{c^2}}}.$$

Moreover, according to this equation the time difference $\Delta t'$ of two events with respect to K' does not in general vanish, even when the time difference Δt of the same events with reference to K vanishes. Pure "space-distance" of two events with respect to K results in "time-distance" of the same events with respect to K'. But the discovery of Minkowski, which was of importance for the formal development of the theory of relativity, does not lie here. It is to be found rather in the fact of his recognition that the four-dimensional space-time continuum of the theory of relativity, in its most essential formal properties, shows a pronounced relationship to the three-dimensional continuum of Euclidean geometrical space.[1] In order to give due prominence to this relationship, however, we must replace the usual time coordinate t by an imaginary magnitude $\sqrt{-1}ct$ proportional to it. Under these conditions, the natural laws satisfying the demands of the (special) theory of relativity assume mathematical forms, in which the time coordinate plays exactly the same role as the three space coordinates. Formally, these four coordinates correspond exactly to the three space coordinates in Euclidean geometry. It must be clear even to the non-mathematician that, as a consequence of this purely formal addition to our knowledge, the theory perforce gained clearness in no mean measure.

These inadequate remarks can give the reader only a vague notion of the important idea contributed by Minkowski. Without it the general theory of relativity, of which the fundamental ideas are developed in the following pages, would perhaps have got no farther than its long clothes. Minkowski's work is doubtless difficult of access to anyone inexperienced in mathematics, but since it is not necessary to have a very exact grasp of this work in order to understand the fundamental ideas of either the special or the general theory of relativity, I shall at present leave it here, and shall revert to it only towards the end of Part II.

---

[1] Cf. the somewhat more detailed discussion in Appendix 2.

II. The General Theory of Relativity

18 SPECIAL AND GENERAL PRINCIPLE OF RELATIVITY

The basal principle, which was the pivot of all our previous considerations, was the *special* principle of relativity, i.e., the principle of the physical relativity of all *uniform* motion. Let us once more analyse its meaning carefully.

It was at all times clear that, from the point of view of the idea it conveys to us, every motion must only be considered as a relative motion. Returning to the illustration we have frequently used of the embankment and the railway carriage, we can express the fact of the motion here taking place in the following two forms, both of which are equally justifiable:

(a) The carriage is in motion relative to the embankment.

(b) The embankment is in motion relative to the carriage.

In (a) the embankment, in (b) the carriage, serves as the body of reference in our statement of the motion taking place. If it is simply a question of detecting or of describing the motion involved, it is in principle immaterial to what reference-body we refer the motion. As already mentioned, this is self-evident, but it must not be confused with the much more comprehensive statement called "the principle of relativity," which we have taken as the basis of our investigations.

The principle we have made use of not only maintains that we may equally well choose the carriage or the embankment as our reference-body for the description of any event (for this, too, is self-evident). Our principle rather asserts what follows: If we formulate the general laws of nature as they are obtained from experience, by making use of

(a) the embankment as reference-body,

(b) the railway carriage as reference-body,

then these general laws of nature (e.g., the laws of mechanics or the law of the propagation of light *in vacuo*) have exactly the same form in both cases. This can also be expressed as follows: For the *physical* description of natural processes, neither of the reference-bodies K, K' is unique (lit. "specially marked out") as compared with the other. Unlike the first, this latter statement need not of necessity hold *a priori*; it is not contained in the conceptions of "motion" and "reference-body" and derivable from them; only *experience* can decide as to its correctness or incorrectness.

Up to the present, however, we have by no means maintained the equivalence of *all* bodies of reference K in connection with the formulation of natural laws. Our course was more on the following lines. In the first place, we started out from the assumption that there exists a reference-body K, whose condition of motion is such that the Galileian law holds with respect to it: A particle left to itself and sufficiently far removed from all other particles moves uniformly in a straight line. With reference to K (Galileian reference-body) the laws of nature were to be as simple as possible. But in addition to K, all bodies of reference K' should be given preference in this sense, and they should be exactly equivalent to K for the formulation of natural laws, provided that they are in a state of *uniform rectilinear and non-rotary motion* with respect to K; all these bodies of reference are to be regarded as Galileian reference-bodies. The validity of the principle of relativity was assumed only for these reference-bodies, but not for others (e.g., those possessing motion of a different kind). In this sense we speak of the *special* principle of relativity, or special theory of relativity.

In contrast to this we wish to understand by the "general principle of relativity" the following statement: All bodies of reference K, K', etc., are equivalent for the description of natural phenomena (formulation of the general laws of nature), whatever may be their state of motion. But before proceeding farther, it ought to be pointed out that this formulation must be replaced later by a more abstract one, for reasons which will become evident at a later stage.

Since the introduction of the special principle of relativity has been justified, every intellect which strives after generalisation must feel the temptation to venture the step towards the general principle of relativity. But a simple and apparently quite reliable consideration seems to suggest that, for the present at any rate, there is little hope of success in such an attempt. Let us imagine ourselves transferred to our old friend the railway carriage, which is travelling at a uniform rate. As long as it is moving uniformly, the occupant of the carriage is not sensible of its motion, and it is for this reason that he can without reluctance interpret the facts of the case as indicating that the carriage is at rest but the embankment in motion. Moreover, according to the special principle of relativity, this interpretation is quite justified also from a physical point of view.

If the motion of the carriage is now changed into a non-uniform motion, as for instance by a powerful application of the brakes, then the occupant of the carriage experiences a correspondingly powerful jerk forwards. The retarded motion is manifested in the mechanical behaviour of bodies relative to the person in the railway carriage. The mechanical behaviour is different from that of the case previously considered, and for this reason it would appear to be impossible that the same mechanical laws hold relatively to the non-uniformly moving carriage, as hold with reference to the carriage when at rest or in uniform motion. At all events it is clear that the Galileian law does not hold with respect to the non-uniformly moving carriage. Because of this, we feel compelled at the present juncture to grant a kind of absolute physical reality to non-uniform motion, in opposition to the general principle of relativity. But in what follows we shall soon see that this conclusion cannot be maintained.

19 THE GRAVITATIONAL FIELD

"If we pick up a stone and then let it go, why does it fall to the ground?" The usual answer to this question is: "Because it is attracted by the earth." Modern physics formulates the answer rather differently for the following reason. As a result of the more careful study of electromagnetic phenomena, we have come to regard action at a distance as a process impossible without the intervention of some intermediary medium. If, for instance, a magnet attracts a piece of iron, we cannot be content to regard this as meaning that the magnet acts directly on the iron through the intermediate empty space, but we are constrained to imagine—after the manner of Faraday—that the magnet always calls into being something physically real in the space around it, that something being what we call a "magnetic field." In its turn this magnetic field operates on the piece of iron, so that the latter strives to move towards the magnet. We shall not discuss here the justification for this incidental conception, which is indeed a somewhat arbitrary one. We shall only mention that with its aid electromagnetic phenomena can be theoretically represented much more satisfactorily than without it, and this applies particularly to the transmission of electromagnetic waves.

The effects of gravitation also are regarded in an analogous manner.

The action of the earth on the stone takes place indirectly. The earth produces in its surroundings a gravitational field, which acts on the stone and produces its motion of fall. As we know from experience, the intensity of the action on a body diminishes according to a quite definite law, as we proceed farther and farther away from the earth. From our point of view this means: The law governing the properties of the gravitational field in space must be a perfectly definite one, in order correctly to represent the diminution of gravitational action with the distance from operative bodies. It is something like this: The body (e.g., the earth) produces a field in its immediate neighbourhood directly; the intensity and direction of the field at points farther removed from the body are thence determined by the law which governs the properties in space of the gravitational fields themselves.

In contrast to electric and magnetic fields, the gravitational field exhibits a most remarkable property, which is of fundamental importance for what follows. Bodies which are moving under the sole influence of a gravitational field receive an acceleration, *which does not in the least depend either on the material or on the physical state of the body.* For instance, a piece of lead and a piece of wood fall in exactly the same manner in a gravitational field (*in vacuo*), when they start off from rest or with the same initial velocity. This law, which holds most accurately, can be expressed in a different form in the light of the following consideration.

According to Newton's law of motion, we have

$$(\text{Force}) = (\text{inertial mass}) \times (\text{acceleration}),$$

where the "inertial mass" is a characteristic constant of the accelerated body. If now gravitation is the cause of the acceleration, we then have

$$(\text{Force}) = (\text{gravitational mass})$$
$$\times (\text{intensity of the gravitational field}),$$

where the "gravitational mass" is likewise a characteristic constant for the body. From these two relations follows:

$$(\text{acceleration}) = \frac{(\text{gravitational mass})}{(\text{inertial mass})}$$
$$\times (\text{intensity of the gravitational field}).$$

If now, as we find from experience, the acceleration is to be independent of the nature and the condition of the body and always the same for a given gravitational field, then the ratio of the gravitational to the inertial mass must likewise be the same for all bodies. By a suitable choice of units we can thus make this ratio equal to unity. We then have the following law: The *gravitational* mass of a body is equal to its *inertial* mass.

It is true that this important law had hitherto been recorded in mechanics, but it had not been *interpreted*. A satisfactory interpretation can be obtained only if we recognise the following fact: *The same* quality of a body manifests itself according to circumstances as "inertia" or as "weight" (lit. "heaviness"). In the following chapter we shall show to what extent this is actually the case, and how this question is connected with the general postulate of relativity.

20 THE EQUALITY OF INERTIAL AND GRAVITATIONAL MASS AS AN ARGUMENT FOR THE GENERAL POSTULATE OF RELATIVITY

We imagine a large portion of empty space, so far removed from stars and other appreciable masses, that we have before us approximately the conditions required by the fundamental law of Galilei. It is then possible to choose a Galileian reference-body for this part of space (world), relative to which points at rest remain at rest and points in motion continue permanently in uniform rectilinear motion. As reference-body let us imagine a spacious chest resembling a room with an observer inside who is equipped with apparatus. Gravitation naturally does not exist for this observer. He must fasten himself with strings to the floor, otherwise the slightest impact against the floor will cause him to rise slowly towards the ceiling of the room.

To the middle of the lid of the chest is fixed externally a hook with rope attached, and now a "being" (what kind of a being is immaterial to us) begins pulling at this with a constant force. The chest together with the observer then begin to move "upwards" with a uniformly accelerated motion. In course of time their velocity will reach unheard-of values—provided that we are viewing all this from another reference-body which is not being pulled with a rope.

But how does the man in the chest regard the process? The acceleration of the chest will be transmitted to him by the reaction of the floor of the chest. He must therefore take up this pressure by means of his legs if he does not wish to be laid out full length on the floor. He is then standing in the chest in exactly the same way as anyone stands in a room of a house on our earth. If he release a body which he previously had in his hand, the acceleration of the chest will no longer be transmitted to this body, and for this reason the body will approach the floor of the chest with an accelerated relative motion. The observer will further convince himself *that the acceleration of the body towards the floor of the chest is always of the same magnitude, whatever kind of body he may happen to use for the experiment.*

Relying on his knowledge of the gravitational field (as it was discussed in the preceding chapter), the man in the chest will thus come to the conclusion that he and the chest are in a gravitational field which is constant with regard to time. Of course he will be puzzled for a moment as to why the chest does not fall, in this gravitational field. Just then, however, he discovers the hook in the middle of the lid of the chest and the rope which is attached to it, and he consequently comes to the conclusion that the chest is suspended at rest in the gravitational field.

Ought we to smile at the man and say that he errs in his conclusion? I do not believe we ought to if we wish to remain consistent; we must rather admit that his mode of grasping the situation violates neither reason nor known mechanical laws. Even though it is being accelerated with respect to the "Galileian space" first considered, we can nevertheless regard the chest as being at rest. We have thus good grounds for extending the principle of relativity to include bodies of reference which are accelerated with respect to each other, and as a result we have gained a powerful argument for a generalised postulate of relativity.

We must note carefully that the possibility of this mode of interpretation rests on the fundamental property of the gravitational field of giving all bodies the same acceleration, or, what comes to the same thing, on the law of the equality of inertial and gravitational mass. If this natural law did not exist, the man in the accelerated chest would not be able to interpret the behaviour of the bodies around him on the supposition of a gravitational field, and he would not be justified on the grounds of experience in supposing his reference-body to be "at rest."

Suppose that the man in the chest fixes a rope to the inner side of the lid, and that he attaches a body to the free end of the rope. The result of this will be to stretch the rope so that it will hang "vertically" downwards. If we ask for an opinion of the cause of tension in the rope, the man in the chest will say: "The suspended body experiences a downward force in the gravitational field, and this is neutralised by the tension of the rope; what determines the magnitude of the tension of the rope is the *gravitational mass* of the suspended body." On the other hand, an observer who is poised freely in space will interpret the condition of things thus: "The rope must perforce take part in the accelerated motion of the chest, and it transmits this motion to the body attached to it. The tension of the rope is just large enough to effect the acceleration of the body. That which determines the magnitude of the tension of the rope is the *inertial mass* of the body." Guided by this example, we see that our extension of the principle of relativity implies the *necessity* of the law of the equality of inertial and gravitational mass. Thus we have obtained a physical interpretation of this law.

From our consideration of the accelerated chest we see that a general theory of relativity must yield important results on the laws of gravitation. In point of fact, the systematic pursuit of the general idea of relativity has supplied the laws satisfied by the gravitational field. Before proceeding farther, however, I must warn the reader against a misconception suggested by these considerations. A gravitational field exists for the man in the chest, despite the fact that there was no such field for the coordinate system first chosen. Now we might easily suppose that the existence of a gravitational field is always only an *apparent* one. We might also think that, regardless of the kind of gravitational field which may be present, we could always choose another reference-body such that *no* gravitational field exists with reference to it. This is by no means true for all gravitational fields, but only for those of quite special form. It is, for instance, impossible to choose a body of reference such that, as judged from it, the gravitational field of the earth (in its entirety) vanishes.

We can now appreciate why that argument is not convincing, which we brought forward against the general principle of relativity at the end of Chapter 18. It is certainly true that the observer in the railway carriage experiences a jerk forwards as a result of the application of the brake, and that he recognises in this the non-uniformity of motion (retardation) of the carriage. But he is compelled by nobody to refer this jerk to a "real" acceleration (retardation) of the carriage. He might also interpret his experience thus: "My body of reference (the carriage) remains permanently at rest. With reference to it, however, there exists (during the period of application of the brakes) a gravitational field which is directed forwards and which is variable with respect to time. Under the influence of this field, the embankment together with the earth moves non-uniformly in such a manner that their original velocity in the backwards direction is continuously reduced."

21 IN WHAT RESPECTS ARE THE FOUNDATIONS OF CLASSICAL MECHANICS AND OF THE SPECIAL THEORY OF RELATIVITY UNSATISFACTORY?

We have already stated several times that classical mechanics starts out from the following law: Material particles sufficiently far removed from other material particles continue to move uniformly in a straight line or continue in a state of rest. We have also repeatedly emphasised that this fundamental law can only be valid for bodies of reference K which possess certain unique states of motion, and which are in uniform translational motion relative to each other. Relative to other reference-bodies K the law is not valid. Both in classical mechanics and in the special theory of relativity we therefore differentiate between reference-bodies K relative to which the recognised "laws of nature" can be said to hold, and reference-bodies K relative to which these laws do not hold.

But no person whose mode of thought is logical can rest satisfied with this condition of things. He asks: "How does it come that certain reference-bodies (or their states of motion) are given priority over other reference-bodies (or their states of motion)? *What is the reason for this preference?*" In order to show clearly what I mean by this question, I shall make use of a comparison.

I am standing in front of a gas range. Standing alongside of each other on the range are two pans so much alike that one may be mistaken for the other. Both are half full of water. I notice that steam is being emitted continuously from the one pan, but not from the other. I am surprised at this, even if I have never seen either a gas range or a pan before. But if I now notice a luminous something of bluish colour under the first pan but not under the other, I cease to be astonished, even if I have never before seen a gas flame. For I can only say that this bluish something will cause the emission of the steam, or at least *possibly* it may do so. If, however, I notice the bluish something in neither case, and if I observe that the one continuously emits steam whilst the other does not, then I shall remain astonished and dissatisfied until I have discovered some circumstance to which I can attribute the different behaviour of the two pans.

Analogously, I seek in vain for a real something in classical mechanics (or in the special theory of relativity) to which I can attribute the different behaviour of bodies considered with respect to the reference-systems

55

K and K'.[1] Newton saw this objection and attempted to invalidate it, but without success. But E. Mach recognised it most clearly of all, and because of this objection he claimed that mechanics must be placed on a new basis. It can only be got rid of by means of a physics which is conformable to the general principle of relativity, since the equations of such a theory hold for every body of reference, whatever may be its state of motion.

---

[1]The objection is of importance more especially when the state of motion of the reference-body is of such a nature that it does not require any external agency for its maintenance, e.g., in the case when the reference-body is rotating uniformly.

22 A FEW INFERENCES FROM THE GENERAL PRINCIPLE OF RELATIVITY

The considerations of Chapter 20 show that the general principle of relativity puts us in a position to derive properties of the gravitational field in a purely theoretical manner. Let us suppose, for instance, that we know the space-time "course" for any natural process whatsoever, as regards the manner in which it takes place in the Galileian domain relative to a Galileian body of reference K. By means of purely theoretical operations (i.e., simply by calculation) we are then able to find how this known natural process appears, as seen from a reference-body K' which is accelerated relatively to K. But since a gravitational field exists with respect to this new body of reference K', our consideration also teaches us how the gravitational field influences the process studied.

For example, we learn that a body which is in a state of uniform rectilinear motion with respect to K (in accordance with the law of Galilei) is executing an accelerated and in general curvilinear motion with respect to the accelerated reference-body K' (chest). This acceleration or curvature corresponds to the influence on the moving body of the gravitational field prevailing relatively to K'. It is known that a gravitational field influences the movement of bodies in this way, so that our consideration supplies us with nothing essentially new.

However, we obtain a new result of fundamental importance when we carry out the analogous consideration for a ray of light. With respect to the Galileian reference-body K, such a ray of light is transmitted rectilinearly with the velocity c. It can easily be shown that the path of the same ray of light is no longer a straight line when we consider it with reference to the accelerated chest (reference-body K'). From this we conclude, *that, in general, rays of light are propagated curvilinearly in gravitational fields.* In two respects this result is of great importance.

In the first place, it can be compared with the reality. Although a detailed examination of the question shows that the curvature of light rays required by the general theory of relativity is only exceedingly small for the gravitational fields at our disposal in practice, its estimated magnitude for light rays passing the sun at grazing incidence is nevertheless 1.7 seconds of arc. This ought to manifest itself in the following way. As seen from the earth, certain fixed stars appear to be in the neighbourhood of the sun, and are thus capable of observation during a total eclipse of the sun. At such

58

times, these stars ought to appear to be displaced outwards from the sun by an amount indicated above, as compared with their apparent position in the sky when the sun is situated at another part of the heavens. The examination of the correctness or otherwise of this deduction is a problem of the greatest importance, the early solution of which is to be expected of astronomers.[1]

In the second place our result shows that, according to the general theory of relativity, the law of the constancy of the velocity of light *in vacuo*, which constitutes one of the two fundamental assumptions in the special theory of relativity and to which we have already frequently referred, cannot claim any unlimited validity. A curvature of rays of light can only take place when the velocity of propagation of light varies with position. Now we might think that as a consequence of this, the special theory of relativity and with it the whole theory of relativity would be laid in the dust. But in reality this is not the case. We can only conclude that the special theory of relativity cannot claim an unlimited domain of validity; its results hold only so long as we are able to disregard the influences of gravitational fields on the phenomena (e.g., of light).

Since it has often been contended by opponents of the theory of relativity that the special theory of relativity is overthrown by the general theory of relativity, it is perhaps advisable to make the facts of the case clearer by means of an appropriate comparison. Before the development of electrodynamics the laws of electrostatics were looked upon as the laws of electricity. At the present time we know that electric fields can be derived correctly from electrostatic considerations only for the case, which is never strictly realised, in which the electrical masses are quite at rest relatively to each other, and to the coordinate system. Should we be justified in saying that for this reason electrostatics is overthrown by the field-equations of Maxwell in electrodynamics? Not in the least. Electrostatics is contained in electrodynamics as a limiting case; the laws of the latter lead directly to those of the former for the case in which the fields are invariable with regard to time. No fairer destiny could be allotted to any physical theory, than that it should of itself point out the way to the introduction of a more comprehensive theory, in which it lives on as a limiting case.

In the example of the transmission of light just dealt with, we have seen that the general theory of relativity enables us to derive theoretically the influence of a gravitational field on the course of natural processes, the laws of which are already known when a gravitational field is absent. But the most attractive problem, to the solution of which the general theory of relativity supplies the key, concerns the investigation of the laws satisfied by the gravitational field itself. Let us consider this for a moment.

We are acquainted with space-time domains which behave (approximately) in a "Galileian" fashion under suitable choice of reference-body,

[1]By means of the star photographs of two expeditions equipped by a Joint Committee of the Royal and Royal Astronomical Societies, the existence of the deflection of light demanded by theory was confirmed during the solar eclipse of 29th May, 1919. (Cf. Appendix 3.)

i.e., domains in which gravitational fields are absent. If we now refer such a domain to a reference-body K' possessing any kind of motion, then relative to K' there exists a gravitational field which is variable with respect to space and time.[2] The character of this field will of course depend on the motion chosen for K'. According to the general theory of relativity, the general law of the gravitational field must be satisfied for all gravitational fields obtainable in this way. Even though by no means all gravitational fields can be produced in this way, yet we may entertain the hope that the general law of gravitation will be derivable from such gravitational fields of a special kind. This hope has been realised in the most beautiful manner. But between the clear vision of this goal and its actual realisation it was necessary to surmount a serious difficulty, and as this lies deep at the root of things, I dare not withhold it from the reader. We require to extend our ideas of the space-time continuum still farther.

---

[2] This follows from a generalisation of the discussion in Chapter 20.

23 Behaviour of Clocks and Measuring-Rods on a Rotating Body of Reference

Hitherto I have purposely refrained from speaking about the physical interpretation of space- and time-data in the case of the general theory of relativity. As a consequence, I am guilty of a certain slovenliness of treatment, which, as we know from the special theory of relativity, is far from being unimportant and pardonable. It is now high time that we remedy this defect; but I would mention at the outset, that this matter lays no small claims on the patience and on the power of abstraction of the reader.

We start off again from quite special cases, which we have frequently used before. Let us consider a space-time domain in which no gravitational field exists relative to a reference-body K whose state of motion has been suitably chosen. K is then a Galileian reference-body as regards the domain considered, and the results of the special theory of relativity hold relative to K. Let us suppose the same domain referred to a second body of reference K', which is rotating uniformly with respect to K. In order to fix our ideas, we shall imagine K' to be in the form of a plane circular disc, which rotates uniformly in its own plane about its centre. An observer who is sitting eccentrically on the disc K' is sensible of a force which acts outwards in a radial direction, and which would be interpreted as an effect of inertia (centrifugal force) by an observer who was at rest with respect to the original reference-body K. But the observer on the disc may regard his disc as a reference-body which is "at rest"; on the basis of the general principle of relativity he is justified in doing this. The force acting on himself, and in fact on all other bodies which are at rest relative to the disc, he regards as the effect of a gravitational field. Nevertheless, the space-distribution of this gravitational field is of a kind that would not be possible on Newton's theory of gravitation.[1] But since the observer believes in the general theory of relativity, this does not disturb him; he is quite in the right when he believes that a general law of gravitation can be formulated—a law which not only explains the motion of the stars correctly, but also the field of force experienced by himself.

The observer performs experiments on his circular disc with clocks and measuring-rods. In doing so, it is his intention to arrive at exact definitions for the signification of time- and space-data with reference to the circular

---

[1] The field disappears at the centre of the disc and increases proportionally to the distance from the centre as we proceed outwards.

disc K', these definitions being based on his observations. What will be his experience in this enterprise?

To start with, he places one of two identically constructed clocks at the centre of the circular disc, and the other on the edge of the disc, so that they are at rest relative to it. We now ask ourselves whether both clocks go at the same rate from the standpoint of the non-rotating Galileian reference-body K. As judged from this body, the clock at the centre of the disc has no velocity, whereas the clock at the edge of the disc is in motion relative to K in consequence of the rotation. According to a result obtained in Chapter 12, it follows that the latter clock goes at a rate permanently slower than that of the clock at the centre of the circular disc, i.e., as observed from K. It is obvious that the same effect would be noted by an observer whom we will imagine sitting alongside his clock at the centre of the circular disc. Thus on our circular disc, or, to make the case more general, in every gravitational field, a clock will go more quickly or less quickly, according to the position in which the clock is situated (at rest). For this reason it is not possible to obtain a reasonable definition of time with the aid of clocks which are arranged at rest with respect to the body of reference. A similar difficulty presents itself when we attempt to apply our earlier definition of simultaneity in such a case, but I do not wish to go any farther into this question.

Moreover, at this stage the definition of the space coordinates also presents insurmountable difficulties. If the observer applies his standard measuring-rod (a rod which is short as compared with the radius of the disc) tangentially to the edge of the disc, then, as judged from the Galileian system, the length of this rod will be less than 1, since, according to Chapter 12, moving bodies suffer a shortening in the direction of the motion. On the other hand, the measuring-rod will not experience a shortening in length, as judged from K, if it is applied to the disc in the direction of the radius. If, then, the observer first measures the circumference of the disc with his measuring-rod and then the diameter of the disc, on dividing the one by the other, he will not obtain as quotient the familiar number $\pi = 3.14\ldots$, but a larger number,[2] whereas of course, for a disc which is at rest with respect to K, this operation would yield π exactly.[3] This proves

---

[2]Throughout this consideration we have to use the Galileian (non-rotating) system K as reference-body, since we may only assume the validity of the results of the special theory of relativity relative to K (relative to K' a gravitational field prevails).

[3]EDITOR'S NOTE: Einstein's assertion that the circumference of the rotating disc will be larger for a stationary observer is incorrect. He erroneously assumed that the measuring-rod along the circumference contracts but the space (along the circumference) does not and therefore more measuring-rods will fit in the space along the circumference and the circumference will be longer (it will contain more measuring-rods) than when at rest. Unfortunately, this erroneous Lorentzian view (that bodies contract but space itself does not) is a common misconception. This misconception can be immediately overcome when it is taken into account that the Lorentz transformations predict that the distance between two points, as measured by a moving observer, will be shorter than the distance between the same points measured by an observer at rest with respect to the points, *no matter whether these points are the end points of a rod or just two points in space.*

The physical meaning of length contraction was made exceedingly clear by Minkowski

that the propositions of Euclidean geometry cannot hold exactly on the rotating disc, nor in general in a gravitational field, at least if we attribute the length 1 to the rod in all positions and in every orientation. Hence the idea of a straight line also loses its meaning. We are therefore not in a position to define exactly the coordinates x, y, z relative to the disc by means of the method used in discussing the special theory, and as long as the coordinates and times of events have not been defined, we cannot assign an exact meaning to the natural laws in which these occur.

Thus all our previous conclusions based on general relativity would appear to be called in question. In reality we must make a subtle detour in order to be able to apply the postulate of general relativity exactly. I shall prepare the reader for this in the following paragraphs.

---

in his ground-breaking lecture *Space and Time* delivered in 1908 (in: Hermann Minkowski, *Space and Time: Minkowski's papers on relativity*, edited by V. Petkov (Minkowski Institute Press, Montreal 2012), p. 116) – Minkowski showed that not only two observers in relative motion have different times but they also have different spaces (forming an angle) and these spaces intersect two parallel worldliness (representing either the end points of a rod or just two points in the space of one of the observers) under different angles; as a result the distance between the points will be different for the two observers.

In 1909 Ehrenfest arrived at the original formulation of the rotating disc problem (Ehrenfest considered a cylinder – P. Ehrenfest, Gleichförmige Rotation starrer Körper und Relativitätstheorie, *Physikalische Zeitschrift*, 1909, 10: 918) "on the basis of Minkowski's ideas" and correctly concluded that "the periphery of the cylinder has to show a contraction compared to its state of rest: $2\pi R' < 2\pi R$."

EUCLIDEAN AND NON-EUCLIDEAN CONTINUUM

The surface of a marble table is spread out in front of me. I can get from any one point on this table to any other point by passing continuously from one point to a "neighbouring" one, and repeating this process a (large) number of times, or, in other words, by going from point to point without executing "jumps." I am sure the reader will appreciate with sufficient clearness what I mean here by "neighbouring" and by "jumps" (if he is not too pedantic). We express this property of the surface by describing the latter as a continuum.

Let us now imagine that a large number of little rods of equal length have been made, their lengths being small compared with the dimensions of the marble slab. When I say they are of equal length, I mean that one can be laid on any other without the ends overlapping. We next lay four of these little rods on the marble slab so that they constitute a quadrilateral figure (a square), the diagonals of which are equally long. To ensure the equality of the diagonals, we make use of a little testing-rod. To this square we add similar ones, each of which has one rod in common with the first. We proceed in like manner with each of these squares until finally the whole marble slab is laid out with squares. The arrangement is such, that each side of a square belongs to two squares and each corner to four squares.

It is a veritable wonder that we can carry out this business without getting into the greatest difficulties. We only need to think of the following. If at any moment three squares meet at a corner, then two sides of the fourth square are already laid, and, as a consequence, the arrangement of the remaining two sides of the square is already completely determined. But I am now no longer able to adjust the quadrilateral so that its diagonals may be equal. If they are equal of their own accord, then this is an especial favour of the marble slab and of the little rods, about which I can only be thankfully surprised. We must need experience many such surprises if the construction is to be successful.

If everything has really gone smoothly, then I say that the points of the marble slab constitute an Euclidean continuum with respect to the little rod, which has been used as a "distance" (line-interval). By choosing one corner of a square as "origin," I can characterise every other corner of a square with reference to this origin by means of two numbers. I only need state how many rods I must pass over when, starting from the origin, I proceed towards the "right" and then "upwards," in order to arrive at the corner of the square under consideration. These two numbers are then the "Cartesian coordinates" of this corner with reference to the "Cartesian coordinate system," which is determined by the arrangement of little rods.

By making use of the following modification of this abstract experiment, we recognise that there must also be cases in which the experiment would be unsuccessful. We shall suppose that the rods "expand" by an amount proportional to the increase of temperature. We heat the central part of the marble slab, but not the periphery, in which case two of our little rods can still be brought into coincidence at every position on the table. But our construction of squares must necessarily come into disorder during the heating, because the little rods on the central region of the table expand, whereas those on the outer part do not.

With reference to our little rods—defined as unit lengths—the marble slab is no longer an Euclidean continuum, and we are also no longer in the position of defining Cartesian coordinates directly with their aid, since the above construction can no longer be carried out. But since there are other things which are not influenced in a similar manner to the little rods (or perhaps not at all) by the temperature of the table, it is possible quite naturally to maintain the point of view that the marble slab is an "Euclidean continuum." This can be done in a satisfactory manner by making a more subtle stipulation about the measurement or the comparison of lengths.

But if rods of every kind (i.e., of every material) were to behave *in the same way* as regards the influence of temperature when they are on the variably heated marble slab, and if we had no other means of detecting the effect of temperature than the geometrical behaviour of our rods in experiments analogous to the one described above, then our best plan would be to assign the distance *one* to two points on the slab, provided that the ends of one of our rods could be made to coincide with these two points; for how else should we define the distance without our proceeding being in the highest measure grossly arbitrary? The method of Cartesian coordinates must then be discarded, and replaced by another which does not assume the validity of Euclidean geometry for rigid bodies.[1] The reader will notice that the situation depicted here corresponds to the one brought about by the general postulate of relativity (Chapter 23).

---

[1]Mathematicians have been confronted with our problem in the following form. If we are given a surface (e.g., an ellipsoid) in Euclidean three-dimensional space, then there exists for this surface a two-dimensional geometry, just as much as for a plane surface. Gauss undertook the task of treating this two-dimensional geometry from first principles, without making use of the fact that the surface belongs to an Euclidean continuum of three dimensions. If we imagine constructions to be made with rigid rods *in the surface* (similar to that above with the marble slab), we should find that different laws hold for these from those resulting on the basis of Euclidean plane geometry. The surface is not an Euclidean continuum with respect to the rods, and we cannot define Cartesian coordinates *in the surface*. Gauss indicated the principles according to which we can treat the geometrical relationships in the surface, and thus pointed out the way to the method of Riemann of treating multi-dimensional, non-Euclidean *continua*. Thus it is that mathematicians long ago solved the formal problems to which we are led by the general postulate of relativity.

25 GAUSSIAN COORDINATES

According to Gauss, this combined analytical and geometrical mode of handling the problem can be arrived at in the following way. We imagine a system of arbitrary curves (see Fig. 4) drawn on the surface of the table. These we designate as u-curves, and we indicate each of them by means of a number. The curves $u = 1$, $u = 2$ and $u = 3$ are drawn in the diagram. Between the curves $u = 1$ and $u = 2$ we must imagine an infinitely large number to be drawn, all of which correspond to real numbers lying between 1 and 2. We have then a system of u-curves, and this "infinitely dense" system covers the whole surface of the table. These u-curves must not intersect each other, and through each point of the surface one and only one curve must pass. Thus a perfectly definite value of u belongs to every point on the surface of the marble slab. In like manner we imagine a system of v-curves drawn on the surface. These satisfy the same conditions as the u-curves, they are

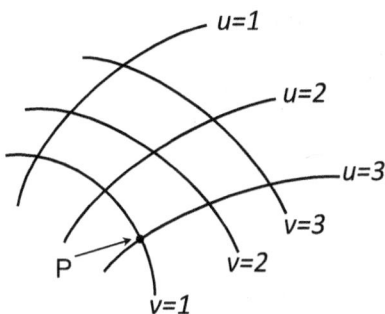

Fig. 4

provided with numbers in a corresponding manner, and they may likewise be of arbitrary shape. It follows that a value of u and a value of v belong to every point on the surface of the table. We call these two numbers the coordinates of the surface of the table (Gaussian coordinates). For example, the point P in the diagram has the Gaussian coordinates $u = 3$, $v = 1$. Two neighbouring points P and P' on the surface then correspond to the coordinates

$$P : \qquad\qquad u, v$$
$$P' : \qquad\qquad u + du, v + dv,$$

where du and dv signify very small numbers. In a similar manner we may indicate the distance (line-interval) between P and P', as measured with a little rod, by means of the very small number ds. Then according to Gauss we have

$$ds^2 = g_{11}\, du^2 + 2g_{12}\, du\, dv + g_{22}\, dv^2,$$

65

where g_{11}, g_{12}, g_{22} are magnitudes which depend in a perfectly definite way on u and v. The magnitudes g_{11}, g_{12} and g_{22} determine the behaviour of the rods relative to the u-curves and v-curves, and thus also relative to the surface of the table. For the case in which the points of the surface considered form an Euclidean continuum with reference to the measuring-rods, but only in this case, it is possible to draw the u-curves and v-curves and to attach numbers to them, in such a manner, that we simply have:

$$ds^2 = du^2 + dv^2.$$

Under these conditions, the u-curves and v-curves are straight lines in the sense of Euclidean geometry, and they are perpendicular to each other. Here the Gaussian coordinates are simply Cartesian ones. It is clear that Gaussian coordinates are nothing more than an association of two sets of numbers with the points of the surface considered, of such a nature that numerical values differing very slightly from each other are associated with neighbouring points "in space."

So far, these considerations hold for a continuum of two dimensions. But the Gaussian method can be applied also to a continuum of three, four or more dimensions. If, for instance, a continuum of four dimensions be supposed available, we may represent it in the following way. With every point of the continuum we associate arbitrarily four numbers, x_1, x_2, x_3, x_4, which are known as "coordinates." Adjacent points correspond to adjacent values of the coordinates. If a distance ds is associated with the adjacent points P and P', this distance being measurable and well-defined from a physical point of view, then the following formula holds:

$$ds^2 = g_{11}\, dx_1{}^2 + 2g_{12}\, dx_1\, dx_2 + \cdots + g_{44}\, dx_4{}^2,$$

where the magnitudes g_{11}, etc., have values which vary with the position in the continuum. Only when the continuum is an Euclidean one is it possible to associate the coordinates $x_1 \ldots x_4$ with the points of the continuum so that we have simply

$$ds^2 = dx_1{}^2 + dx_2{}^2 + dx_3{}^2 + dx_4{}^2.$$

In this case relations hold in the four-dimensional continuum which are analogous to those holding in our three-dimensional measurements.

However, the Gauss treatment for ds^2 which we have given above is not always possible. It is only possible when sufficiently small regions of the continuum under consideration may be regarded as Euclidean continua.

For example, this obviously holds in the case of the marble slab of the table and local variation of temperature. The temperature is practically constant for a small part of the slab, and thus the geometrical behaviour of the rods is *almost* as it ought to be according to the rules of Euclidean geometry. Hence the imperfections of the construction of squares in the previous chapter do not show themselves clearly until this construction is extended over a considerable portion of the surface of the table.

We can sum this up as follows: Gauss invented a method for the mathematical treatment of continua in general, in which "size-relations" ("distances" between neighbouring points) are defined. To every point of a continuum are assigned as many numbers (Gaussian coordinates) as the continuum has dimensions. This is done in such a way, that only one meaning can be attached to the assignment, and that numbers (Gaussian coordinates) which differ by an indefinitely small amount are assigned to adjacent points. The Gaussian coordinate system is a logical generalisation of the Cartesian coordinate system. It is also applicable to non-Euclidean continua, but only when, with respect to the defined "size" or "distance," small parts of the continuum under consideration behave more nearly like an Euclidean system, the smaller the part of the continuum under our notice.

26 The Space-Time Continuum of the Special Theory of Relativity Considered as an Euclidean Continuum

We are now in a position to formulate more exactly the idea of Minkowski, which was only vaguely indicated in Chapter 17. In accordance with the special theory of relativity, certain coordinate systems are given preference for the description of the four-dimensional, space-time continuum. We called these "Galileian coordinate systems." For these systems, the four coordinates x, y, z, t, which determine an event or—in other words—a point of the four-dimensional continuum, are defined physically in a simple manner, as set forth in detail in the first part of this book. For the transition from one Galileian system to another, which is moving uniformly with reference to the first, the equations of the Lorentz transformation are valid. These last form the basis for the derivation of deductions from the special theory of relativity, and in themselves they are nothing more than the expression of the universal validity of the law of transmission of light for all Galileian systems of reference.

Minkowski found that the Lorentz transformations satisfy the following simple conditions. Let us consider two neighbouring events, the relative position of which in the four-dimensional continuum is given with respect to a Galileian reference-body K by the space coordinate differences dx, dy, dz and the time-difference dt. With reference to a second Galileian system we shall suppose that the corresponding differences for these two events are dx', dy', dz', dt'. Then these magnitudes always fulfil the condition[1]

$$dx^2 + dy^2 + dz^2 - c^2\,dt^2 = dx'^2 + dy'^2 + dz'^2 - c^2\,dt'^2.$$

The validity of the Lorentz transformation follows from this condition. We can express this as follows: The magnitude

$$ds^2 = dx^2 + dy^2 + dz^2 - c^2\,dt^2,$$

which belongs to two adjacent points of the four-dimensional space-time continuum, has the same value for all selected (Galileian) reference-bodies.

---

[1] Cf. Appendices 1 and 2. The relations which are derived there for the coordinates themselves are valid also for coordinate *differences*, and thus also for coordinate differentials (indefinitely small differences).

If we replace x, y, z, $\sqrt{-1}\,ct$, by x_1, x_2, x_3, x_4, we also obtain the result that

$$ds^2 = dx_1{}^2 + dx_2{}^2 + dx_3{}^2 + dx_4{}^2$$

is independent of the choice of the body of reference. We call the magnitude ds the "distance" apart of the two events or four-dimensional points.

Thus, if we choose as time-variable the imaginary variable $\sqrt{-1}\,ct$ instead of the real quantity t, we can regard the space-time continuum—in accordance with the special theory of relativity—as an "Euclidean" four-dimensional continuum, a result which follows from the considerations of the preceding chapter.

27 THE SPACE-TIME CONTINUUM OF THE GENERAL THEORY OF RELATIVITY IS NOT AN EUCLIDEAN CONTINUUM

In the first part of this book we were able to make use of space-time co-ordinates which allowed of a simple and direct physical interpretation, and which, according to Chapter 26, can be regarded as four-dimensional Cartesian coordinates. This was possible on the basis of the law of the constancy of the velocity of light. But according to Chapter 22, the general theory of relativity cannot retain this law. On the contrary, we arrived at the result that according to this latter theory the velocity of light must always depend on the coordinates when a gravitational field is present. In connection with a specific illustration in Chapter 23, we found that the presence of a gravitational field invalidates the definition of the coordinates and the time, which led us to our objective in the special theory of relativity.

In view of the results of these considerations we are led to the conviction that, according to the general principle of relativity, the space-time continuum cannot be regarded as an Euclidean one, but that here we have the general case, corresponding to the marble slab with local variations of temperature, and with which we made acquaintance as an example of a two-dimensional continuum. Just as it was there impossible to construct a Cartesian coordinate system from equal rods, so here it is impossible to build up a system (reference-body) from rigid bodies and clocks, which shall be of such a nature that measuring-rods and clocks, arranged rigidly with respect to one another, shall indicate position and time directly. Such was the essence of the difficulty with which we were confronted in Chapter 23.

But the considerations of Chapters 25 and 26 show us the way to surmount this difficulty. We refer the four-dimensional space-time continuum in an arbitrary manner to Gaussian coordinates. We assign to every point of the continuum (event) four numbers, x_1, x_2, x_3, x_4 (coordinates), which have not the least direct physical significance, but only serve the purpose of numbering the points of the continuum in a definite but arbitrary manner. This arrangement does not even need to be of such a kind that we must regard x_1, x_2, x_3 as "space" coordinates and x_4 as a "time" coordinate.

The reader may think that such a description of the world would be quite inadequate. What does it mean to assign to an event the particular coordinates x_1, x_2, x_3, x_4, if in themselves these coordinates have no significance? More careful consideration shows, however, that this anxiety is unfounded. Let us consider, for instance, a material point with any kind of motion. If this point had only a momentary existence without duration, then it would be described in space-time by a single system of values x_1, x_2, x_3, x_4. Thus its permanent existence must be characterised by an infinitely large number of such systems of values, the coordinate values of which are so close together as to give continuity; corresponding to the material point, we thus have a (uni-dimensional) line in the four-dimensional continuum. In the same way, any such lines in our continuum correspond to many points in motion. The only statements having regard to these points which can claim a physical existence are in reality the statements about their encounters. In our mathematical treatment, such an encounter is expressed in the fact that the two lines which represent the motions of the points in question have a particular system of coordinate values, x_1, x_2, x_3, x_4, in common. After mature consideration the reader will doubtless admit that in reality such encounters constitute the only actual evidence of a time-space nature with which we meet in physical statements.

When we were describing the motion of a material point relative to a body of reference, we stated nothing more than the encounters of this point with particular points of the reference-body. We can also determine the corresponding values of the time by the observation of encounters of the body with clocks, in conjunction with the observation of the encounter of the hands of clocks with particular points on the dials. It is just the same in the case of space-measurements by means of measuring-rods, as a little consideration will show.

The following statements hold generally: Every physical description resolves itself into a number of statements, each of which refers to the space-time coincidence of two events A and B. In terms of Gaussian coordinates, every such statement is expressed by the agreement of their four coordinates x_1, x_2, x_3, x_4. Thus in reality, the description of the time-space continuum by means of Gaussian coordinates completely replaces the description with the aid of a body of reference, without suffering from the defects of the latter mode of description; it is not tied down to the Euclidean character of the continuum which has to be represented.

28   Exact Formulation of the General Principle of Relativity

We are now in a position to replace the provisional formulation of the general principle of relativity given in Chapter 18 by an exact formulation. The form there used, "All bodies of reference K, K', etc., are equivalent for the description of natural phenomena (formulation of the general laws of nature), whatever may be their state of motion," cannot be maintained, because the use of rigid reference-bodies, in the sense of the method followed in the special theory of relativity, is in general not possible in space-time description. The Gaussian coordinate system has to take the place of the body of reference. The following statement corresponds to the fundamental idea of the general principle of relativity: "*All Gaussian coordinate systems are essentially equivalent for the formulation of the general laws of nature.*"

We can state this general principle of relativity in still another form, which renders it yet more clearly intelligible than it is when in the form of the natural extension of the special principle of relativity. According to the special theory of relativity, the equations which express the general laws of nature pass over into equations of the same form when, by making use of the Lorentz transformation, we replace the space-time variables x, y, z, t, of a (Galileian) reference-body K by the space-time variables x', y', z', t', of a new reference-body K'. According to the general theory of relativity, on the other hand, by application of *arbitrary substitutions* of the Gaussian variables x_1, x_2, x_3, x_4, the equations must pass over into equations of the same form; for every transformation (not only the Lorentz transformation) corresponds to the transition of one Gaussian coordinate system into another.

If we desire to adhere to our "old-time" three-dimensional view of things, then we can characterise the development which is being undergone by the fundamental idea of the general theory of relativity as follows: The special theory of relativity has reference to Galileian domains, i.e., to those in which no gravitational field exists. In this connection a Galileian reference-body serves as body of reference, i.e., a rigid body the state of motion of which is so chosen that the Galileian law of the uniform rectilinear motion of "isolated" material points holds relatively to it.

Certain considerations suggest that we should refer the same Galileian domains to *non-Galileian* reference-bodies also. A gravitational field of a special kind is then present with respect to these bodies (cf. Chapters 20 and 23).

In gravitational fields there are no such things as rigid bodies with Euclidean properties; thus the fictitious rigid body of reference is of no avail in the general theory of relativity. The motion of clocks is also influenced by gravitational fields, and in such a way that a physical definition of time which is made directly with the aid of clocks has by no means the same degree of plausibility as in the special theory of relativity.

For this reason non-rigid reference-bodies are used, which are as a whole not only moving in any way whatsoever, but which also suffer alterations in form *ad lib.* during their motion. Clocks, for which the law of motion is of any kind, however irregular, serve for the definition of time. We have to imagine each of these clocks fixed at a point on the non-rigid reference-body. These clocks satisfy only the one condition, that the "readings" which are observed simultaneously on adjacent clocks (in space) differ from each other by an indefinitely small amount. This non-rigid reference-body, which might appropriately be termed a "reference-mollusk," is in the main equivalent to a Gaussian four-dimensional coordinate system chosen arbitrarily. That which gives the "mollusk" a certain comprehensibility as compared with the Gaussian coordinate system is the (really unjustified) formal retention of the separate existence of the space coordinates as opposed to the time coordinate. Every point on the mollusk is treated as a space-point, and every material point which is at rest relatively to it as at rest, so long as the mollusk is considered as reference-body. The general principle of relativity requires that all these mollusks can be used as reference-bodies with equal right and equal success in the formulation of the general laws of nature; the laws themselves must be quite independent of the choice of mollusk.

The great power possessed by the general principle of relativity lies in the comprehensive limitation which is imposed on the laws of nature in consequence of what we have seen above.

29 THE SOLUTION OF THE PROBLEM OF GRAVITATION ON THE BASIS OF THE GENERAL PRINCIPLE OF RELATIVITY

If the reader has followed all our previous considerations, he will have no further difficulty in understanding the methods leading to the solution of the problem of gravitation.

We start off from a consideration of a Galileian domain, i.e., a domain in which there is no gravitational field relative to the Galileian reference-body K. The behaviour of measuring-rods and clocks with reference to K is known from the special theory of relativity, likewise the behaviour of "isolated" material points; the latter move uniformly and in straight lines.

Now let us refer this domain to a random Gaussian coordinate system or to a "mollusk" as reference-body K'. Then with respect to K' there is a gravitational field G (of a particular kind). We learn the behaviour of measuring-rods and clocks and also of freely-moving material points with reference to K' simply by mathematical transformation. We interpret this behaviour as the behaviour of measuring-rods, clocks and material points under the influence of the gravitational field G. Hereupon we introduce a hypothesis: that the influence of the gravitational field on measuring-rods, clocks and freely-moving material points continues to take place according to the same laws, even in the case when the prevailing gravitational field is *not* derivable from the Galileian special case, simply by means of a transformation of coordinates.

The next step is to investigate the space-time behaviour of the gravitational field G, which was derived from the Galileian special case simply by transformation of the coordinates. This behaviour is formulated in a law, which is always valid, no matter how the reference-body (mollusk) used in the description may be chosen.

This law is not yet the *general* law of the gravitational field, since the gravitational field under consideration is of a special kind. In order to find out the general law-of-field of gravitation we still require to obtain a generalisation of the law as found above. This can be obtained without caprice, however, by taking into consideration the following demands:

(a) The required generalisation must likewise satisfy the general postulate of relativity.

(b) If there is any matter in the domain under consideration, only its inertial mass, and thus according to Chapter 15 only its energy is of importance for its effect in exciting a field.

(c) Gravitational field and matter together must satisfy the law of the conservation of energy (and of impulse).

Finally, the general principle of relativity permits us to determine the influence of the gravitational field on the course of all those processes which take place according to known laws when a gravitational field is absent, i.e., which have already been fitted into the frame of the special theory of relativity. In this connection we proceed in principle according to the method which has already been explained for measuring-rods, clocks and freely-moving material points.

The theory of gravitation derived in this way from the general postulate of relativity excels not only in its beauty; nor in removing the defect attaching to classical mechanics which was brought to light in Chapter 21; nor in interpreting the empirical law of the equality of inertial and gravitational mass; but it has also already explained a result of observation in astronomy, against which classical mechanics is powerless.

If we confine the application of the theory to the case where the gravitational fields can be regarded as being weak, and in which all masses move with respect to the coordinate system with velocities which are small compared with the velocity of light, we then obtain as a first approximation the Newtonian theory. Thus the latter theory is obtained here without any particular assumption, whereas Newton had to introduce the hypothesis that the force of attraction between mutually attracting material points is inversely proportional to the square of the distance between them. If we increase the accuracy of the calculation, deviations from the theory of Newton make their appearance, practically all of which must nevertheless escape the test of observation owing to their smallness.

We must draw attention here to one of these deviations. According to Newton's theory, a planet moves round the sun in an ellipse, which would permanently maintain its position with respect to the fixed stars, if we could disregard the motion of the fixed stars themselves and the action of the other planets under consideration. Thus, if we correct the observed motion of the planets for these two influences, and if Newton's theory be strictly correct, we ought to obtain for the orbit of the planet an ellipse, which is fixed with reference to the fixed stars. This deduction, which can be tested with great accuracy, has been confirmed for all the planets save one, with the precision that is capable of being obtained by the delicacy of observation attainable at the present time. The sole exception is Mercury, the planet which lies nearest the sun. Since the time of Leverrier, it has been known that the ellipse corresponding to the orbit of Mercury, after it has been corrected for the influences mentioned above, is not stationary with respect to the fixed stars, but that it rotates exceedingly slowly in the plane of the orbit and in the sense of the orbital motion. The value obtained for this rotary movement of the orbital ellipse was 43 seconds of arc per century, an amount ensured to be correct to within a few seconds of arc. This effect can be explained by means of classical mechanics only on the assumption of hypotheses which have little probability, and which were devised solely for this purpose.

On the basis of the general theory of relativity, it is found that the ellipse of every planet round the sun must necessarily rotate in the manner indicated above; that for all the planets, with the exception of Mercury, this rotation is too small to be detected with the delicacy of observation possible at the present time; but that in the case of Mercury it must amount to 43 seconds of arc per century, a result which is strictly in agreement with observation.

Apart from this one, it has hitherto been possible to make only two deductions from the theory which admit of being tested by observation, to wit, the curvature of light rays by the gravitational field of the sun,[1] and a displacement of the spectral lines of light reaching us from large stars, as compared with the corresponding lines for light produced in an analogous manner terrestrially (i.e., by the same kind of atom).[2,3] These two deductions from the theory have both been confirmed.

---

[1] Observed by Eddington and others in 1919. (Cf. Appendix 3)

[2] Established by Adams in 1924. (Cf. p. 98).

[3] EDITOR'S NOTE: See the EDITOR'S NOTE on p. 130.

III. Consideration on the Universe as a Whole

30   COSMOLOGICAL DIFFICULTIES OF NEWTON'S THEORY

Apart from the difficulty discussed in Chapter 21, there is a second fundamental difficulty attending classical celestial mechanics, which, to the best of my knowledge, was first discussed in detail by the astronomer Seeliger. If we ponder over the question as to how the universe, considered as a whole, is to be regarded, the first answer that suggests itself to us is surely this: As regards space (and time) the universe is infinite. There are stars everywhere, so that the density of matter, although very variable in detail, is nevertheless on the average everywhere the same. In other words: However far we might travel through space, we should find everywhere an attenuated swarm of fixed stars of approximately the same kind and density.

This view is not in harmony with the theory of Newton. The latter theory rather requires that the universe should have a kind of centre in which the density of the stars is a maximum, and that as we proceed outwards from this centre the group-density of the stars should diminish, until finally, at great distances, it is succeeded by an infinite region of emptiness. The stellar universe ought to be a finite island in the infinite ocean of space.[1]

This conception is in itself not very satisfactory. It is still less satisfactory because it leads to the result that the light emitted by the stars and also individual stars of the stellar system are perpetually passing out into infinite space, never to return, and without ever again coming into interaction with other objects of nature. Such a finite material universe would be destined to become gradually but systematically impoverished.

In order to escape this dilemma, Seeliger suggested a modification of Newton's law, in which he assumes that for great distances the force of attraction between two masses diminishes more rapidly than would result from the inverse square law. In this way it is possible for the mean density of matter to be constant everywhere, even to infinity, without infinitely large gravitational fields being produced. We thus free ourselves from the distasteful conception that the material universe ought to possess something

---

[1] *Proof*—According to the theory of Newton, the number of "lines of force" which come from infinity and terminate in a mass m is proportional to the mass m. If, on the average, the mass-density ρ_0 is constant throughout the universe, then a sphere of volume V will enclose the average mass $\rho_0 V$. Thus the number of lines of force passing through the surface F of the sphere into its interior is proportional to $\rho_0 V$. For unit area of the surface of the sphere the number of lines of force which enters the sphere is thus proportional to $\rho_0 V/F$ or to $\rho_0 R$. Hence the intensity of the field at the surface would ultimately become infinite with increasing radius R of the sphere, which is impossible.

of the nature of a centre. Of course we purchase our emancipation from the fundamental difficulties mentioned, at the cost of a modification and complication of Newton's law which has neither empirical nor theoretical foundation. We can imagine innumerable laws which would serve the same purpose, without our being able to state a reason why one of them is to be preferred to the others; for any one of these laws would be founded just as little on more general theoretical principles as is the law of Newton.

31 THE POSSIBILITY OF A "FINITE" AND YET "UNBOUNDED" UNIVERSE

But speculations on the structure of the universe also move in quite another direction. The development of non-Euclidean geometry led to the recognition of the fact, that we can cast doubt on the *infiniteness* of our space without coming into conflict with the laws of thought or with experience (Riemann, Helmholtz). These questions have already been treated in detail and with unsurpassable lucidity by Helmholtz and Poincaré, whereas I can only touch on them briefly here.

In the first place, we imagine an existence in two-dimensional space. Flat beings with flat implements, and in particular flat rigid measuring-rods, are free to move in a *plane*. For them nothing exists outside of this plane: that which they observe to happen to themselves and to their flat "things" is the all-inclusive reality of their plane. In particular, the constructions of plane Euclidean geometry can be carried out by means of the rods, e.g., the lattice construction, considered in Chapter 24. In contrast to ours, the universe of these beings is two-dimensional; but, like ours, it extends to infinity. In their universe there is room for an infinite number of identical squares made up of rods, i.e., its volume (surface) is infinite. If these beings say their universe is "plane," there is sense in the statement, because they mean that they can perform the constructions of plane Euclidean geometry with their rods. In this connection the individual rods always represent the same distance, independently of their position.

Let us consider now a second two-dimensional existence, but this time on a spherical surface instead of on a plane. The flat beings with their measuring-rods and other objects fit exactly on this surface and they are unable to leave it. Their whole universe of observation extends exclusively over the surface of the sphere. Are these beings able to regard the geometry of their universe as being plane geometry and their rods withal as the realisation of "distance"? They cannot do this. For if they attempt to realise a straight line, they will obtain a curve, which we "three-dimensional beings" designate as a great circle, i.e., a self-contained line of definite finite length, which can be measured up by means of a measuring-rod. Similarly, this universe has a finite area that can be compared with the area of a square constructed with rods. The great charm resulting from this consideration lies in the recognition of the fact that *the universe of these beings is finite and yet has no limits.*

But the spherical-surface beings do not need to go on a world-tour in order to perceive that they are not living in an Euclidean universe. They can convince themselves of this on every part of their "world," provided they do not use too small a piece of it. Starting from a point, they draw "straight lines" (arcs of circles as judged in three-dimensional space) of equal length in all directions. They will call the line joining the free ends of these lines a "circle." For a plane surface, the ratio of the circumference of a circle to its diameter, both lengths being measured with the same rod, is, according to Euclidean geometry of the plane, equal to a constant value π, which is independent of the diameter of the circle. On their spherical surface our flat beings would find for this ratio the value

$$\pi = \frac{\sin\left(\dfrac{r}{R}\right)}{\left(\dfrac{r}{R}\right)},$$

i.e., a smaller value than π, the difference being the more considerable, the greater is the radius of the circle in comparison with the radius R of the "world-sphere." By means of this relation the spherical beings can determine the radius of their universe ("world"), even when only a relatively small part of their world-sphere is available for their measurements. But if this part is very small indeed, they will no longer be able to demonstrate that they are on a spherical "world" and not on an Euclidean plane, for a small part of a spherical surface differs only slightly from a piece of a plane of the same size.

Thus if the spherical-surface beings are living on a planet of which the solar system occupies only a negligibly small part of the spherical universe, they have no means of determining whether they are living in a finite or in an infinite universe, because the "piece of universe" to which they have access is in both cases practically plane, or Euclidean. It follows directly from this discussion, that for our sphere-beings the circumference of a circle first increases with the radius until the "circumference of the universe" is reached, and that it thenceforward gradually decreases to zero for still further increasing values of the radius. During this process the area of the circle continues to increase more and more, until finally it becomes equal to the total area of the whole "world-sphere."

Perhaps the reader will wonder why we have placed our "beings" on a sphere rather than on another closed surface. But this choice has its justification in the fact that, of all closed surfaces, the sphere is unique in possessing the property that all points on it are equivalent. I admit that the ratio of the circumference c of a circle to its radius r depends on r, but for a given value of r it is the same for all points of the "world-sphere"; in other words, the "world-sphere" is a "surface of constant curvature."

To this two-dimensional sphere-universe there is a three-dimensional analogy, namely, the three-dimensional spherical space which was discovered by Riemann. Its points are likewise all equivalent. It possesses a finite volume, which is determined by its "radius" $(2\pi^2 R^3)$. Is it possible to imagine a spherical space? To imagine a space means nothing else than that we imagine an epitome of our "space" experience, i.e., of experience that we can have in the movement of "rigid" bodies. In this sense we *can* imagine a spherical space.

Suppose we draw lines or stretch strings in all directions from a point, and mark off from each of these the distance r with a measuring-rod. All the free end-points of these lengths lie on a spherical surface. We can specially measure up the area (F) of this surface by means of a square made up of measuring-rods. If the universe is Euclidean, then $F = 4\pi r^2$; if it is spherical, then F is always less than $4\pi r^2$. With increasing values of r, F increases from zero up to a maximum value which is determined by the "world-radius," but for still further increasing values of r, the area gradually diminishes to zero. At first, the straight lines which radiate from the starting point diverge farther and farther from one another, but later they approach each other, and finally they run together again at a "counter-point" to the starting point. Under such conditions they have traversed the whole spherical space. It is easily seen that the three-dimensional spherical space is quite analogous to the two-dimensional spherical surface. It is finite (i.e., of finite volume), and has no bounds.

It may be mentioned that there is yet another kind of curved space: "elliptical space." It can be regarded as a curved space in which the two "counter-points" are identical (indistinguishable from each other). An elliptical universe can thus be considered to some extent as a curved universe possessing central symmetry.

It follows from what has been said, that closed spaces without limits are conceivable. From amongst these, the spherical space (and the elliptical) excels in its simplicity, since all points on it are equivalent. As a result of this discussion, a most interesting question arises for astronomers and physicists, and that is whether the universe in which we live is infinite, or whether it is finite in the manner of the spherical universe. Our experience is far from being sufficient to enable us to answer this question. But the general theory of relativity permits of our answering it with a moderate degree of certainty, and in this connection the difficulty mentioned in Chapter 30 finds its solution.

32 THE STRUCTURE OF SPACE ACCORDING TO THE GENERAL THEORY OF RELATIVITY

According to the general theory of relativity, the geometrical properties of space are not independent, but they are determined by matter. Thus we can draw conclusions about the geometrical structure of the universe only if we base our considerations on the state of the matter as being something that is known. We know from experience that, for a suitably chosen coordinate system, the velocities of the stars are small as compared with the velocity of transmission of light. We can thus as a rough approximation arrive at a conclusion as to the nature of the universe as a whole, if we treat the matter as being at rest.

We already know from our previous discussion that the behaviour of measuring-rods and clocks is influenced by gravitational fields, i.e., by the distribution of matter. This in itself is sufficient to exclude the possibility of the exact validity of Euclidean geometry in our universe. But it is conceivable that our universe differs only slightly from an Euclidean one, and this notion seems all the more probable, since calculations show that the metrics of surrounding space is influenced only to an exceedingly small extent by masses even of the magnitude of our sun. We might imagine that, as regards geometry, our universe behaves analogously to a surface which is irregularly curved in its individual parts, but which nowhere departs appreciably from a plane: something like the rippled surface of a lake. Such a universe might fittingly be called a quasi-Euclidean universe. As regards its space it would be infinite. But calculation shows that in a quasi-Euclidean universe the average density of matter would necessarily be *nil*. Thus such a universe could not be inhabited by matter everywhere; it would present to us that unsatisfactory picture which we portrayed in Chapter 30.

If we are to have in the universe an average density of matter which differs from zero, however small may be that difference, then the universe cannot be quasi-Euclidean. On the contrary, the results of calculation indicate that if matter be distributed uniformly, the universe would necessarily be spherical (or elliptical). Since in reality the detailed distribution of matter is not uniform, the real universe will deviate in individual parts from the spherical, i.e., the universe will be quasi-spherical. But it will be necessarily finite. In fact, the theory supplies us with a simple connection[1] between the space-expanse of the universe and the average density of matter in it.

---

[1]For the "radius" R of the universe we obtain the equation

$$R^2 = \frac{2}{\kappa\rho}.$$

The use of the C.G.S. system in this equation gives $2/\kappa = 1.08 \cdot 10^{27}$; ρ is the average density of the matter and κ is a constant connected with the Newtonian constant of gravitation.

APPENDICES

APPENDIX 1
Simple Derivation of the Lorentz Transformation (Supplementary to Chapter 11)

For the relative orientation of the coordinate systems indicated in Fig. 2, the x-axes of both systems permanently coincide. In the present case we can divide the problem into parts by considering first only events which are localised on the x-axis. Any such event is represented with respect to the coordinate system K by the abscissa x and the time t, and with respect to the system K' by the abscissa x' and the time t'. We require to find x' and t' when x and t are given.

A light-signal, which is proceeding along the positive axis of x, is transmitted according to the equation

$$x = ct$$

or

$$x - ct = 0. \tag{1}$$

Since the same light-signal has to be transmitted relative to K' with the velocity c, the propagation relative to the system K' will be represented by the analogous formula

$$x' - ct' = 0. \tag{2}$$

Those space-time points (events) which satisfy (1) must also satisfy (2). Obviously this will be the case when the relation

$$(x' - ct') = \lambda(x - ct) \tag{3}$$

is fulfilled in general, where λ indicates a constant; for, according to (3), the disappearance of $(x - ct)$ involves the disappearance of $(x' - ct')$.

If we apply quite similar considerations to light rays which are being transmitted along the negative x-axis, we obtain the condition

$$(x' + ct') = \mu(x + ct). \tag{4}$$

By adding (or subtracting) equations (3) and (4), and introducing for convenience the constants a and b in place of the constants λ and μ, where

$$a = \frac{\lambda + \mu}{2}$$

and

$$b = \frac{\lambda - \mu}{2},$$

we obtain the equations

$$\left.\begin{array}{l} x' = ax - bct \\ ct' = act - bx \end{array}\right\} \tag{5}$$

We should thus have the solution of our problem, if the constants a and b were known. These result from the following discussion.

For the origin of K' we have permanently $x' = 0$, and hence according to the first of the equations (5)

$$x = \frac{bc}{a}t.$$

If we call v the velocity with which the origin of K' is moving relative to K, we then have

$$v = \frac{bc}{a}. \tag{6}$$

The same value v can be obtained from equation (5), if we calculate the velocity of another point of K' relative to K, or the velocity (directed towards the negative x-axis) of a point of K with respect to K'. In short, we can designate v as the relative velocity of the two systems.

Furthermore, the principle of relativity teaches us that, as judged from K, the length of a unit measuring-rod which is at rest with reference to K' must be exactly the same as the length, as judged from K', of a unit measuring-rod which is at rest relative to K. In order to see how the points of the x'-axis appear as viewed from K, we only require to take a "snapshot" of K' from K; this means that we have to insert a particular value of t (time of K), e.g., $t = 0$. For this value of t we then obtain from the first of the equations (5)

$$x' = ax.$$

Two points of the x'-axis which are separated by the distance $\Delta x' = 1$ when measured in the K' system are thus separated in our instantaneous photograph by the distance

$$\Delta x = \frac{1}{a}. \tag{7}$$

But if the snapshot be taken from $K'(t' = 0)$, and if we eliminate t from the equations (5), taking into account the expression (6), we obtain

$$x' = a\left(1 - \frac{v^2}{c^2}\right)x.$$

From this we conclude that two points on the x-axis and separated by the distance 1 (relative to K) will be represented on our snapshot by the distance

$$\Delta x' = a\left(1 - \frac{v^2}{c^2}\right). \tag{7a}$$

But from what has been said, the two snapshots must be identical; hence Δx in (7) must be equal to $\Delta x'$ in (7a), so that we obtain

$$a^2 = \frac{1}{1 - \dfrac{v^2}{c^2}}. \tag{7b}$$

The equations (6) and (7b) determine the constants a and b. By inserting the values of these constants in (5), we obtain the first and the fourth of the equations given in Chapter 11.

$$\left.\begin{aligned} x' &= \frac{x - vt}{\sqrt{1 - \dfrac{v^2}{c^2}}} \\[2em] t' &= \frac{t - \dfrac{v}{c^2} x}{\sqrt{1 - \dfrac{v^2}{c^2}}} \end{aligned}\right\} \tag{8}$$

Thus we have obtained the Lorentz transformation for events on the x-axis. It satisfies the condition

$$x'^2 - c^2 t'^2 = x^2 - c^2 t^2. \tag{8a}$$

The extension of this result, to include events which take place outside the x-axis, is obtained by retaining equations (8) and supplementing them by the relations

$$\left.\begin{aligned} y' &= y \\ z' &= z \end{aligned}\right\} \tag{9}$$

In this way we satisfy the postulate of the constancy of the velocity of light *in vacuo* for rays of light of arbitrary direction, both for the system K and for the system K'. This may be shown in the following manner.

We suppose a light-signal sent out from the origin of K at the time $t = 0$. It will be propagated according to the equation

$$r = \sqrt{x^2 + y^2 + z^2} = ct,$$

or, if we square this equation, according to the equation

$$x^2 + y^2 + z^2 - c^2 t^2 = 0. \tag{10}$$

It is required by the law of propagation of light, in conjunction with the postulate of relativity, that the transmission of the signal in question should take place—as judged from K'—in accordance with the corresponding formula

$$r' = ct',$$

or,

$$x'^2 + y'^2 + z'^2 - c^2 t'^2 = 0. \tag{10a}$$

In order that equation (10a) may be a consequence of equation (10), we must have

$$x'^2 + y'^2 + z'^2 - c^2 t'^2 = \sigma(x^2 + y^2 + z^2 - c^2 t^2). \qquad (11)$$

Since equation (8a) must hold for points on the x-axis, we thus have $\sigma = 1$. It is easily seen that the Lorentz transformation really satisfies equation (11) for $\sigma = 1$; for (11) is a consequence of (8a) and (9), and hence also of (8) and (9). We have thus derived the Lorentz transformation.

The Lorentz transformation represented by (8) and (9) still requires to be generalised. Obviously it is immaterial whether the axes of K' be chosen so that they are spatially parallel to those of K. It is also not essential that the velocity of translation of K' with respect to K should be in the direction of the x-axis. A simple consideration shows that we are able to construct the Lorentz transformation in this general sense from two kinds of transformations, viz. from Lorentz transformations in the special sense and from purely spatial transformations, which corresponds to the replacement of the rectangular coordinate system by a new system with its axes pointing in other directions.

Mathematically, we can characterise the generalised Lorentz transformation thus:

It expresses x', y', z', t', in terms of linear homogeneous functions of x, y, z, t, of such a kind that the relation

$$x'^2 + y'^2 + z'^2 - c^2 t'^2 = x^2 + y^2 + z^2 - c^2 t^2 \qquad (11a)$$

is satisfied identically. That is to say: If we substitute their expressions in x, y, z, t, in place of x', y', z', t', on the left-hand side, then the left-hand side of (11a) agrees with the right-hand side.

Appendix 2
Minkowski's Four-Dimensional Space ("World") (Supplementary to Chapter 17)

We can characterise the Lorentz transformation still more simply if we introduce the imaginary $\sqrt{-1}ct$ in place of t, as time-variable. If, in accordance with this, we insert

$$x_1 = x$$
$$x_2 = y$$
$$x_3 = z$$
$$x_4 = \sqrt{-1}ct,$$

and similarly for the accented system K', then the condition which is identically satisfied by the transformation can be expressed thus:

$$x_1'^2 + x_2'^2 + x_3'^2 + x_4'^2 = x_1{}^2 + x_2{}^2 + x_3{}^2 + x_4{}^2. \tag{12}$$

That is, by the afore-mentioned choice of "coordinates," (11a) is transformed into this equation.

We see from (12) that the imaginary time coordinate x_4 enters into the condition of transformation in exactly the same way as the space coordinates x_1, x_2, x_3. It is due to this fact that, according to the theory of relativity, the "time" x_4 enters into natural laws in the same form as the space coordinates x_1, x_2, x_3.

A four-dimensional continuum described by the "coordinates" x_1, x_2, x_3, x_4, was called "world" by Minkowski, who also termed a point-event a "world-point." From a "happening" in three-dimensional space, physics becomes, as it were, an "existence" in the four-dimensional "world."

This four-dimensional "world" bears a close similarity to the three-dimensional "space" of (Euclidean) analytical geometry. If we introduce into the latter a new Cartesian coordinate system (x_1', x_2', x_3') with the same origin, then x_1', x_2', x_3', are linear homogeneous functions of x_1, x_2, x_3, which identically satisfy the equation

$$x_1'^2 + x_2'^2 + x_3'^2 = x_1{}^2 + x_2{}^2 + x_3{}^2.$$

The analogy with (12) is a complete one. We can regard Minkowski's "world" in a formal manner as a four-dimensional Euclidean space (with imaginary time coordinate); the Lorentz transformation corresponds to a "rotation" of the coordinate system in the four-dimensional "world."

Appendix 3
The Experimental Confirmation of the General Theory of Relativity

From a systematic theoretical point of view, we may imagine the process of evolution of an empirical science to be a continuous process of induction. Theories are evolved and are expressed in short compass as statements of a large number of individual observations in the form of empirical laws, from which the general laws can be ascertained by comparison. Regarded in this way, the development of a science bears some resemblance to the compilation of a classified catalogue. It is, as it were, a purely empirical enterprise.

But this point of view by no means embraces the whole of the actual process; for it slurs over the important part played by intuition and deductive thought in the development of an exact science. As soon as a science has emerged from its initial stages, theoretical advances are no longer achieved merely by a process of arrangement. Guided by empirical data, the investigator rather develops a system of thought which, in general, is built up logically from a small number of fundamental assumptions, the so-called axioms. We call such a system of thought a *theory*. The theory finds the justification for its existence in the fact that it correlates a large number of single observations, and it is just here that the "truth" of the theory lies.

Corresponding to the same complex of empirical data, there may be several theories, which differ from one another to a considerable extent. But as regards the deductions from the theories which are capable of being tested, the agreement between the theories may be so complete, that it becomes difficult to find any deductions in which the two theories differ from each other. As an example, a case of general interest is available in the province of biology, in the Darwinian theory of the development of species by selection in the struggle for existence, and in the theory of development which is based on the hypothesis of the hereditary transmission of acquired characters.

We have another instance of far-reaching agreement between the deductions from two theories in Newtonian mechanics on the one hand, and the general theory of relativity on the other. This agreement goes so far, that up to the present we have been able to find only a few deductions from the general theory of relativity which are capable of investigation, and to which the physics of pre-relativity days does not also lead, and this despite the profound difference in the fundamental assumptions of the two theories. In what follows, we shall again consider these important deductions, and we shall also discuss the empirical evidence appertaining to them which has hitherto been obtained.

(a) Motion of the Perihelion of Mercury

According to Newtonian mechanics and Newton's law of gravitation, a planet which is revolving round the sun would describe an ellipse round the latter, or, more correctly, round the common centre of gravity of the sun and the planet. In such a system, the sun, or the common centre of gravity, lies in one of the foci of the orbital ellipse in such a manner that, in the course of a planet-year, the distance sun-planet grows from a minimum to a maximum, and then decreases again to a minimum. If instead of Newton's law we insert a somewhat different law of attraction into the calculation, we find that, according to this new law, the motion would still take place in such a manner that the distance sun-planet exhibits periodic variations; but in this case the angle described by the line joining sun and planet during such a period (from perihelion—closest proximity to the sun—to perihelion) would differ from 360^0. The line of the orbit would not then be a closed one, but in the course of time it would fill up an annular part of the orbital plane, viz. between the circle of least and the circle of greatest distance of the planet from the sun.

According also to the general theory of relativity, which differs of course from the theory of Newton, a small variation from the Newton-Kepler motion of a planet in its orbit should take place, and in such a way, that the angle described by the radius sun-planet between one perihelion and the next should exceed that corresponding to one complete revolution by an amount given by

$$+\frac{24\,\pi^3 a^2}{T^2 c^2 (1 - e^2)}.$$

(*N.B.*—One complete revolution corresponds to the angle 2π in the absolute angular measure customary in physics, and the above expression gives the amount by which the radius sun-planet exceeds this angle during the interval between one perihelion and the next.) In this expression a represents the major semi-axis of the ellipse, e its eccentricity, c the velocity of light, and T the period of revolution of the planet. Our result may also be stated as follows: According to the general theory of relativity, the major axis of the ellipse rotates round the sun in the same sense as the orbital motion of the planet. Theory requires that this rotation should amount to 43 seconds of arc per century for the planet Mercury, but for the other planets of our solar system its magnitude should be so small that it would necessarily escape detection.[1]

In point of fact, astronomers have found that the theory of Newton does not suffice to calculate the observed motion of Mercury with an exactness corresponding to that of the delicacy of observation attainable at the present time. After taking account of all the disturbing influences exerted on Mercury by the remaining planets, it was found (Leverrier—1859—and Newcomb—1895) that an unexplained perihelial movement of the orbit of Mercury remained over, the amount of which does not differ sensibly from the above-mentioned +43 seconds of arc per century. The uncertainty of the empirical result amounts to a few seconds only.

---

[1] Especially since the next planet Venus has an orbit that is almost an exact circle, which makes it more difficult to locate the perihelion with precision.

(b) Deflection of Light by a Gravitational Field

In Chapter 22 it has been already mentioned that, according to the general theory of relativity, a ray of light will experience a curvature of its path when passing through a gravitational field, this curvature being similar to that experienced by the path of a body which is projected through a gravitational field. As a result of this theory, we should expect that a ray of light which is passing close to a heavenly body would be deviated towards the latter. For a ray of light which passes the sun at a distance of Δ sun-radii from its centre, the angle of deflection (α) should amount to

$$\alpha = \frac{1.7 \text{ seconds of arc}}{\Delta}.$$

It may be added that, according to the theory, half of this deflection is produced by the Newtonian field of attraction of the sun, and the other half by the geometrical modification ("curvature") of space caused by the sun.

This result admits of an experimental test by means of the photographic registration of stars during a total eclipse of the sun. The only reason why we must wait for a total eclipse is because at every other time the atmosphere is so strongly illuminated by the light from the sun that the stars situated near the sun's disc are invisible. The predicted effect can be seen clearly from the accompanying diagram. If the sun (S) were not present, a star which is practically infinitely distant would be seen in the direction D_1, as observed from the earth. But as a consequence of

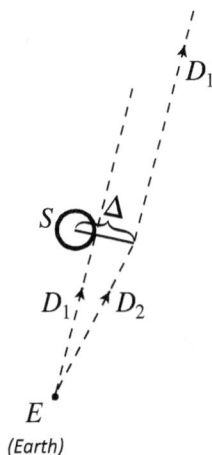

Fig. 5

the deflection of light from the star by the sun, the star will be seen in the direction D_2, i.e., at a somewhat greater distance from the centre of the sun than corresponds to its real position.

In practice, the question is tested in the following way. The stars in the neighbourhood of the sun are photographed during a solar eclipse. In addition, a second photograph of the same stars is taken when the sun is situated at another position in the sky, i.e., a few months earlier or later. As compared with the standard photograph, the positions of the stars on the eclipse-photograph ought to appear displaced radially outwards (away from the centre of the sun) by an amount corresponding to the angle α.

We are indebted to the Royal Society and to the Royal Astronomical Society for the investigation of this important deduction. Undaunted by the war and by difficulties of both a material and a psychological nature aroused by the war, these societies equipped two expeditions—to Sobral (Brazil), and to the island of Principe (West Africa)—and sent several of Britain's most celebrated astronomers (Eddington, Cottingham, Crommelin, Davidson), in order to obtain photographs of the solar eclipse of 29th May, 1919. The relative discrepancies to be expected between the stellar photographs obtained during the eclipse and the comparison photographs amounted to a few hundredths of a millimetre only. Thus great accuracy was necessary in making the adjustments required for the taking of the photographs, and in their subsequent measurement.

The results of the measurements confirmed the theory in a thoroughly satisfactory manner. The rectangular components of the observed and of the calculated deviations of the stars (in seconds of arc) are set forth in the following table of results:

Number of the Star.	First Co-ordinate		Second Co-ordinate	
	Observed.	Calculated.	Observed.	Calculated.
11	−0.19	−0.22	+0.16	+0.02
5	+0.29	+0.31	−0.46	−0.43
4	+0.11	+0.10	+0.83	+0.74
3	+0.20	+0.12	+1.00	+0.87
6	+0.10	+0.04	+0.57	+0.40
10	−0.08	+0.09	+0.35	+0.32
2	+0.95	+0.85	−0.27	−0.09

(c) Displacement of Spectral Lines towards the Red

In Chapter 23 it has been shown that in a system K' which is in rotation with regard to a Galileian system K, clocks of identical construction, and which are considered at rest with respect to the rotating reference-body, go at rates which are dependent on the positions of the clocks. We shall now examine this dependence quantitatively. A clock, which is situated at a distance r from the centre of the disc, has a velocity relative to K which is given by

$$v = \omega r,$$

where ω represents the angular velocity of rotation of the disc K' with respect to K. If ν_0 represents the number of ticks of the clock per unit time ("rate" of the clock) relative to K when the clock is at rest, then the "rate" of the clock (ν) when it is moving relative to K with a velocity v, but at rest with respect to the disc, will, in accordance with Chapter 12, be given by

$$\nu = \nu_0 \sqrt{1 - \frac{v^2}{c^2}},$$

or with sufficient accuracy by

$$\nu = \nu_0 \left(1 - \frac{1}{2}\frac{v^2}{c^2}\right).$$

This expression may also be stated in the following form:

$$\nu = \nu_0 \left(1 - \frac{1}{c^2} \frac{\omega^2 r^2}{2}\right).$$

If we represent the difference of potential of the centrifugal force between the position of the clock and the centre of the disc by ϕ, i.e., the work, considered negatively, which must be performed on the unit of mass against the centrifugal force in order to transport it from the position of the clock on the rotating disc to the centre of the disc, then we have

$$\phi = -\frac{\omega^2 r^2}{2}.$$

From this it follows that

$$\nu = \nu_0 \left(1 + \frac{\phi}{c^2}\right).$$

In the first place, we see from this expression that two clocks of identical construction will go at different rates when situated at different distances from the centre of the disc. This result is also valid from the standpoint of an observer who is rotating with the disc.

Now, as judged from the disc, the latter is in a gravitational field of potential ϕ, hence the result we have obtained will hold quite generally for gravitational fields. Furthermore, we can regard an atom which is emitting spectral lines as a clock, so that the following statement will hold:

An atom absorbs or emits light of a frequency which is dependent on the potential of the gravitational field in which it is situated.

The frequency of an atom situated on the surface of a heavenly body will be somewhat less than the frequency of an atom of the same element which is situated in free space (or on the surface of a smaller celestial body). Now $\phi = -K\frac{M}{r}$, where K is Newton's constant of gravitation, and M is the mass of the heavenly body. Thus a displacement towards the red ought to take place for spectral lines produced at the surface of stars as compared with the spectral lines of the same element produced at the surface of the earth, the amount of this displacement being

$$\frac{\nu_0 - \nu}{\nu_0} = \frac{K}{c^2} \frac{M}{r}.$$

For the sun, the displacement towards the red predicted by theory amounts to about two millionths of the wave-length. A trustworthy calculation is not possible in the case of the stars, because in general neither the mass M nor the radius r are known.

It is an open question whether or not this effect exists, and at the present time (1920) astronomers are working with great zeal towards the solution. Owing to the smallness of the effect in the case of the sun, it is difficult to form an opinion as to its existence. Whereas Grebe and Bachem

(Bonn), as a result of their own measurements and those of Evershed and Schwarzschild on the cyanogen bands, have placed the existence of the effect almost beyond doubt, other investigators, particularly St. John, have been led to the opposite opinion in consequence of their measurements.

Mean displacements of lines towards the less refrangible end of the spectrum are certainly revealed by statistical investigations of the fixed stars; but up to the present the examination of the available data does not allow of any definite decision being arrived at, as to whether or not these displacements are to be referred in reality to the effect of gravitation. The results of observation have been collected together, and discussed in detail from the standpoint of the question which has been engaging our attention here, in a paper by E. Freundlich entitled "Zur Prüfung der allgemeinen Relativitäts-Theorie" (*Die Naturwissenschaften*, 1919, No. 35, p. 520: Julius Springer, Berlin).

At all events, a definite decision will be reached during the next few years. If the displacement of spectral lines towards the red by the gravitational potential does not exist, then the general theory of relativity will be untenable. On the other hand, if the cause of the displacement of spectral lines be definitely traced to the gravitational potential, then the study of this displacement will furnish us with important information as to the mass of the heavenly bodies.

NOTE:[2] —The displacement of spectral lines towards the red end of the spectrum was definitely established by Adams in 1924, by observations on the dense companion of Sirius, for which the effect is about thirty times greater than for the sun.

R. W. L.

---

[2]EDITOR'S NOTE: See the EDITOR'S NOTE on p. 130.

APPENDIX 4

The Structure of Space According to the General Theory of Relativity (Supplementary to Chapter 32)

Since the publication of the first edition of this little book, our knowledge about the structure of space in the large ("cosmological problem") has had an important development, which ought to be mentioned even in a popular presentation of the subject. My original considerations on the subject were based on two hypotheses:

(1) There exists an average density of matter in the whole of space which is everywhere the same and different from zero.

(2) The magnitude ("radius") of space is independent of time.

Both these hypotheses proved to be consistent, according to the general theory of relativity, but only after a hypothetical term was added to the field equations, a term which was not required by the theory as such nor did it seem natural from a theoretical point of view ("cosmological term of the field equations").

Hypothesis (2) appeared unavoidable to me at the time, since I thought that one would get into bottomless speculations if one departed from it.

However, already in the twenties, the Russian mathematician Friedmann showed that a different hypothesis was natural from a purely theoretical point of view. He realized that it was possible to preserve hypothesis (1) without introducing the less natural cosmological term into the field equations of gravitation, if one was ready to drop hypothesis (2). Namely, the original field equations admit a solution in which the "world radius" depends on time (expanding space). In that sense one can say, according to Friedmann, that the theory demands an expansion of space.

A few years later Hubble showed, by a special investigation of the extra-galactic nebulae ("milky ways"), that the spectral lines emitted showed a red shift which increased regularly with the distance of the nebulae. This can be interpreted in regard to our present knowledge only in the sense of Doppler's principle, as an expansive motion of the system of stars in the large – as required, according to Friedmann, by the field equations of gravitation. Hubble's discovery can, therefore, be considered to some extent as a confirmation of the theory.

There does arise, however, a strange difficulty. The interpretation of the galactic line-shift discovered by Hubble as an expansion (which can hardly be doubted from a theoretical point of view), leads to an origin of this expansion which lies "only" about 10^9 years ago, while physical astronomy makes it appear likely that the development of individual stars and systems of stars takes considerably longer. It is in no way known how this incongruity is to be overcome.

I further want to remark that the theory of expanding space, together with the empirical data of astronomy, permit no decision to be reached about the finite or infinite character of (three-dimensional) space, while the original "static" hypothesis of space yielded the closure (finiteness) of space.

APPENDIX 5
Relativity and the Problem of Space[1]

It is characteristic of Newtonian physics that it has to ascribe independent and real existence to space and time as well as to matter, for in Newton's law of motion the idea of acceleration appears. But in this theory, acceleration can only denote "acceleration with respect to space". Newton's space must thus be thought of as "at rest", or at least as "unaccelerated", in order that one can consider the acceleration, which appears in the law of motion, as being a magnitude with any meaning. Much the same holds with time, which of course likewise enters into the concept of acceleration. Newton himself and his most critical contemporaries felt it to be disturbing that one had to ascribe physical reality both to space itself as well as to its state of motion; but there was at that time no other alternative, if one wished to ascribe to mechanics a clear meaning.

It is indeed an exacting requirement to have to ascribe physical reality to space in general, and especially to empty space. Time and again since remotest times philosophers have resisted such a presumption. Descartes argued somewhat on these lines: space is identical with extension, but extension is connected with bodies; thus there is no space without bodies and hence no empty space. The weakness of this argument lies primarily in what follows. It is certainly true that the concept extension owes its origin to our experiences of laying out or bringing into contact solid bodies. But from this it cannot be concluded that the concept of extension may not be justified in cases which have not themselves given rise to the formation of this concept. Such an enlargement of concepts can be justified indirectly by its value for the comprehension of empirical results. The assertion that extension is confined to bodies is therefore of itself certainly unfounded. We shall see later, however, that the general theory of relativity confirms Descartes' conception in a roundabout way. What brought Descartes to his remarkably attractive view was certainly the feeling that, without compelling necessity, one ought not to ascribe reality to a thing like space, which is not capable of being "directly experienced".[2]

The psychological origin of the idea of space, or of the necessity for it, is far from being so obvious as it may appear to be on the basis of our customary habit of thought. The old geometers deal with conceptual objects (straight line, point, surface), but not really with space as such, as was done later in analytical geometry. The idea of space, however, is suggested by certain primitive experiences. Suppose that a box has been

---

[1]As with the original translation of this book in 1920, my old friend Emeritus Professor S. R. Milner, F.R.S. has again given me the benefit of his unique experience in this field, by reading the translation of this new appendix and making numerous suggestions for improvement. I am deeply grateful to him and to Professor A. G. Walker of the Mathematics Department of Liverpool University, who also read this appendix and offered various helpful suggestions. R. W. L.

[2]This expression is to be taken *cum grano salis*.

constructed. Objects can be arranged in a certain way inside the box, so that it becomes full. The possibility of such arrangements is a property of the material object "box", something that is given with the box, the "space enclosed" by the box. This is something which is different for different boxes, something that is thought quite naturally as being independent of whether or not, at any moment, there are any objects at all in the box. When there are no objects in the box, its space appears to be "empty".

So far, our concept of space has been associated with the box. It turns out, however, that the storage possibilities that make up the box-space are independent of the thickness of the walls of the box. Cannot this thickness be reduced to zero, without the "space" being lost as a result? The naturalness of such a limiting process is obvious, and now there remains for our thought the space without the box, a self-evident thing, yet it appears to be so unreal if we forget the origin of this concept. One can understand that it was repugnant to Descartes to consider space as independent of material objects, a thing that might exist without matter.[3] (At the same time, this does not prevent him from treating space as a fundamental concept in his analytical geometry.) The drawing of attention to the vacuum in a mercury barometer has certainly disarmed the last of the Cartesians. But it is not to be denied that, even at this primitive stage, something unsatisfactory clings to the concept of space, or to space thought of as an independent real thing.

The ways in which bodies can be packed into space (e.g. the box) are the subject of three-dimensional Euclidean geometry, whose axiomatic structure readily deceives us into forgetting that it refers to realisable situations.

If now the concept of space is formed in the manner outlined above, and following on from experience about the "filling" of the box, then this space is primarily a *bounded* space. This limitation does not appear to be essential, however, for apparently a larger box can always be introduced to enclose the smaller one. In this way space appears as something unbounded.

I shall not consider here how the concepts of the three-dimensional and the Euclidean nature of space can be traced back to relatively primitive experiences. Rather, I shall consider first of all from other points of view the role of the concept of space in the development of physical thought.

When a smaller box s is situated, relatively at rest, inside the hollow space of a larger box S, then the hollow space of s is a part of the hollow space of S, and the same "space", which contains both of them, belongs to each of the boxes. When s is in motion with respect to S, however, the concept is less simple. One is then inclined to think that s encloses always the same space, but a variable part of the space S. It then becomes necessary to apportion to each box its particular space, not thought of as

---

[3] Kant's attempt to remove the embarrassment by denial of the objectivity of space can, however, hardly be taken seriously. The possibilities of packing inherent in the inside space of a box are objective in the same sense as the box itself, and as the objects which can be packed inside it.

bounded, and to assume that these two spaces are in motion with respect to each other.

Before one has become aware of this complication, space appears as an unbounded medium or container in which material objects swim around. But it must now be remembered that there is an infinite number of spaces, which are in motion with respect to each other. The concept of space as something existing objectively and independent of things belongs to pre-scientific thought, but not so the idea of the existence of an infinite number of spaces in motion relatively to each other. This latter idea is indeed logically unavoidable, but is far from having played a considerable role even in scientific thought.

But what about the psychological origin of the concept of time? This concept is undoubtedly associated with the fact of "calling to mind," as well as with the differentiation between sense experiences and the recollection of these. Of itself it is doubtful whether the differentiation between sense experience and recollection (or simple re-presentation) is something psychologically directly given to us. Everyone has experienced that he has been in doubt whether he has actually experienced something with his senses or has simply dreamt about it. Probably the ability to discriminate between these alternatives first comes about as the result of an activity of the mind creating order.

An experience is associated with a "recollection", and it is considered as being "earlier" in comparison with "present experiences". This is a conceptual ordering principle for recollected experiences, and the possibility of its accomplishment gives rise to the subjective concept of time, i.e., that concept of time which refers to the arrangement of the experiences of the individual.

What do we mean by rendering objective the concept of time? Let us consider an example. A person A ("I") has the experience "it is lightning". At the same time the person A also experiences such a behaviour of the person B as brings the behaviour of B into relation with his own experience "it is lightning". Thus it comes about that A associates with B the experience "it is lightning". For the person A the idea arises that other persons also participate in the experience "it is lightning". "It is lightning" is now no longer interpreted as an exclusively personal experience, but as an experience of other persons (or eventually only as a "potential experience"). In this way arises the interpretation that "it is lightning", which originally entered into the consciousness as an "experience", is now also interpreted as an (objective) "event". It is just the sum total of all events that we mean when we speak of the "real external world".

We have seen that we feel ourselves impelled to ascribe a temporal arrangement to our experiences, somewhat as follows. If β is later than α and γ later than β, then γ is also later than α ("sequence of experiences"). Now what is the position in this respect with the "events" which we have associated with the experiences? At first sight it seems obvious to assume that a temporal arrangement of events exists which agrees with the temporal arrangement of the experiences. In general, and unconsciously this was done, until sceptical doubts made themselves felt.[4] In order to arrive at the idea of an objective world, an additional constructive concept still is necessary: the event is localised not only in time, but also in space.

In the previous paragraphs we have attempted to describe how the concepts space, time and event can be put psychologically into relation with experiences. Considered logically, they are free creations of the human intelligence, tools of thought, which are to serve the purpose of bringing experiences into relation with each other, so that in this way they can be better surveyed. The attempt to become conscious of the empirical sources of these fundamental concepts should show to what extent we are actually bound to these concepts. In this way we become aware of our freedom, of which, in case of necessity, it is always a difficult matter to make sensible use.

We still have something essential to add to this sketch concerning the psychological origin of the concepts space-time-event (we will call them more briefly "space-like", in contrast to concepts from the psychological sphere). We have linked up the concept of space with experiences using boxes and the arrangement of material objects in them. Thus this formation of concepts already presupposes the concept of material objects (e.g. "boxes"). In the same way persons, who had to be introduced for the formation of an objective concept of time, also play the role of material objects in this connection. It appears to me, therefore, that the formation of the concept of the material object must precede our concepts of time and space.

All these space-like concepts already belong to pre-scientific thought, along with concepts like pain, goal, purpose, etc. from the field of psychology. Now it is characteristic of thought in physics, as of thought in natural science generally, that it endeavours in principle to make do with "space-like" concepts *alone*, and strives to express with their aid all relations having the form of laws. The physicist seeks to reduce colours and tones to vibrations, the physiologist thought and pain to nerve processes, in such a way that the psychical element as such is eliminated from the causal nexus of existence, and thus nowhere occurs as an independent link in the causal associations. It is no doubt this attitude, which considers the comprehension of all relations by the exclusive use of only "space-like" concepts as being possible in principle, that is at the present time understood by the term "materialism" (since "matter" has lost its role as a fundamental concept).

Why is it necessary to drag down from the Olympian fields of Plato the fundamental ideas of thought in natural science, and to attempt to reveal their earthly lineage? Answer: In order to free these ideas from the taboo attached to them, and thus to achieve greater freedom in the formation of ideas or concepts. It is to the immortal credit of D. Hume and E. Mach

---

[4]For example, the order of experiences in time obtained by acoustical means can differ from the temporal order gained visually, so that one cannot simply identify the time sequence of events with the time sequence of experiences.

that they, above all others, introduced this critical conception.

Science has taken over from pre-scientific thought the concepts space, time, and material object (with the important special case "solid body"), and has modified them and rendered them more precise. Its first significant accomplishment was the development of Euclidean geometry, whose axiomatic formulation must not be allowed to blind us to its empirical origin (the possibilities of laying out or juxtaposing solid bodies). In particular, the three-dimensional nature of space as well as its Euclidean character are of empirical origin (it can be wholly filled by like constituted "cubes").

The subtlety of the concept of space was enhanced by the discovery that there exist no completely rigid bodies. All bodies are elastically deformable and alter in volume with change in temperature. The structures, whose possible congruences are to be described by Euclidean geometry, cannot therefore be represented apart from physical concepts. But since physics after all must make use of geometry in the establishment of its concepts, the empirical content of geometry can be stated and tested only in the framework of the whole of physics.

In this connection atomistics must also be borne in mind, and its conception of finite divisibility; for spaces of sub-atomic extension cannot be measured up. Atomistics also compels us to give up, in principle, the idea of sharply and statically defined bounding surfaces of solid bodies. Strictly speaking, there are no *precise* laws, even in the macro-region, for the possible configurations of solid bodies touching each other.

In spite of this, no one thought of giving up the concept of space, for it appeared indispensable in the eminently satisfactory whole system of natural science. Mach, in the nineteenth century, was the only one who thought seriously of an elimination of the concept of space, in that he sought to replace it by the notion of the totality of the instantaneous distances between all material points. (He made this attempt in order to arrive at a satisfactory understanding of inertia).

The field

In Newtonian mechanics, space and time play a dual role. First, they play the part of carrier or frame for things that happen in physics, in reference to which events are described by the space coordinates and the time. In principle, matter is thought of as consisting of "material points", the motions of which constitute physical happening. When matter is thought of as being continuous, this is done as it were provisionally in those cases where one does not wish to or cannot describe the discrete structure. In this case small parts (elements of volume) of the matter are treated similarly to material points, at least in so far as we are concerned merely with motions and not with occurrences which, at the moment, it is not possible or serves no useful purpose to attribute to motions (e.g. temperature changes, chemical processes). The second role of space and time was that of being an "inertial system". From all conceivable systems of reference, inertial systems were considered to be advantageous in that, with respect to them, the law of inertia claimed validity.

In this, the essential thing is that "physical reality", thought of as being independent of the subjects experiencing it, was conceived as consisting, at least in principle, of space and time on one hand, and of permanently existing material points, moving with respect to space and time, on the other. The idea of the independent existence of space and time can be expressed drastically in this way: If matter were to disappear, space and time alone would remain behind (as a kind of stage for physical happening).

The surmounting of this standpoint resulted from a development which, in the first place, appeared to have nothing to do with the problem of space-time, namely, the appearance of the *concept of field* and its final claim to replace, in principle, the idea of a particle (material point). In the framework of classical physics, the concept of field appeared as an auxiliary concept, in cases in which matter was treated as a continuum. For example, in the consideration of the heat conduction in a solid body, the state of the body is described by giving the temperature at every point of the body for every definite time. Mathematically, this means that the temperature T is represented as a mathematical expression (function) of the space coordinates and the time t (Temperature field). The law of heat conduction is represented as a local relation (differential equation), which embraces all special cases of the conduction of heat. The temperature is here a simple example of the concept of field. This is a quantity (or a complex of quantities), which is a function of the coordinates and the time. Another example is the description of the motion of a liquid. At every point there exists at any time a velocity, which is quantitatively described by its three "components" with respect to the axes of a coordinate system (vector). The components of the velocity at a point (field components), here also, are functions of the coordinates (x, y, z) and the time (t).

It is characteristic of the fields mentioned that they occur only within a ponderable mass; they serve only to describe a state of this matter. In accordance with the historical development of the field concept, where no matter was available there could also exist no field. But in the first quarter of the nineteenth century it was shown that the phenomena of the interference and motion of light could be explained with astonishing clearness when light was regarded as a wave-field, completely analogous to the mechanical vibration field in an elastic solid body. It was thus felt necessary to introduce a field, that could also exist in "empty space" in the absence of ponderable matter.

This state of affairs created a paradoxical situation, because, in accordance with its origin, the field concept appeared to be restricted to the description of states in the inside of a ponderable body. This seemed to be all the more certain, inasmuch as the conviction was held that every field is to be regarded as a state capable of mechanical interpretation, and this presupposed the presence of matter. One thus felt compelled, even in the space which had hitherto been regarded as empty, to assume everywhere the existence of a form of matter, which was called "aether".

The emancipation of the field concept from the assumption of its association with a mechanical carrier finds a place among the psychologically most interesting events in the development of physical thought. During the second half of the nineteenth century, in connection with the researches of Faraday and Maxwell, it became more and more clear that the description of electromagnetic processes in terms of field was vastly superior to a treatment on the basis of the mechanical concepts of material points. By the introduction of the field concept in electrodynamics, Maxwell succeeded in predicting the existence of electromagnetic waves, the essential identity of which with light waves could not be doubted, because of the equality of their velocity of propagation. As a result of this, optics was, in principle, absorbed by electrodynamics. *One* psychological effect of this immense success was that the field concept, as opposed to the mechanistic framework of classical physics, gradually won greater independence.

Nevertheless, it was at first taken for granted that electromagnetic fields had to be interpreted as states of the aether, and it was zealously sought to explain these states as mechanical ones. But as these efforts always met with frustration, science gradually became accustomed to the idea of renouncing such a mechanical interpretation. Nevertheless, the conviction still remained that electromagnetic fields must be states of the aether, and this was the position at the turn of the century.

The aether-theory brought with it the question: How does the aether behave from the mechanical point of view with respect to ponderable bodies? Does it take part in the motions of the bodies, or do its parts remain at rest relatively to each other? Many ingenious experiments were undertaken to decide this question. The following important facts should be mentioned in this connection: the "aberration" of the fixed stars in consequence of the annual motion of the earth, and the "Doppler effect", i.e. the influence of the relative motion of the fixed stars on the frequency of the light reaching us from them, for known frequencies of emission. The results of all these facts and experiments, except for one, the Michelson-Morley experiment, were explained by H. A. Lorentz on the assumption that the aether does not take part in the motions of ponderable bodies, and that the parts of the aether have no relative motions at all with respect to each other. Thus the aether appeared, as it were, as the embodiment of a space absolutely at rest. But the investigation of Lorentz accomplished still more. It explained all the electromagnetic and optical processes within ponderable bodies known at that time, on the assumption that the influence of ponderable matter on the electric field—and conversely—is due solely to the fact that the constituent particles of matter carry electrical charges, which share the motion of the particles. Concerning the experiment of Michelson and Morley, H. A. Lorentz showed that the result obtained at least does not contradict the theory of an aether at rest.[5]

---

[5] EDITOR'S NOTE: Although the Michelson-Morley experiment can be explained by the aether theory, it turned out that that theory is wrong – e.g., the theory of an aether

108

In spite of all these beautiful successes the state of the theory was not yet wholly satisfactory, and for the following reasons. Classical mechanics, of which it could not be doubted that it holds with a close degree of approximation, teaches the equivalence of all inertial systems or inertial "spaces" for the formulation of natural laws, i.e. the invariance of natural laws with respect to the transition from one inertial system to another. Electromagnetic and optical *experiments* taught the same thing with considerable accuracy. But the foundation of electromagnetic *theory* taught that a particular inertial system must be given preference, namely that of the luminiferous aether at rest. This view of the theoretical foundation was much too unsatisfactory. Was there no modification that, like classical mechanics, would uphold the equivalence of inertial systems (special principle of relativity)?

The answer to this question is the special theory of relativity. This takes over from the theory of Maxwell-Lorentz the assumption of the constancy of the velocity of light in empty space. In order to bring this into harmony with the equivalence of inertial systems (special principle of relativity), the idea of the absolute character of simultaneity must be given up; in addition, the Lorentz transformations for the time and the space coordinates follow for the transition from one inertial system to another. The whole content of the special theory of relativity is included in the postulate: The laws of Nature are invariant with respect to the Lorentz transformations. The important thing of this requirement lies in the fact that it limits the possible natural laws in a definite manner.

What is the position of the special theory of relativity in regard to the problem of space? In the first place we must guard against the opinion that the four-dimensionality of reality has been newly introduced for the first time by this theory. Even in classical physics the event is localised by four numbers, three spatial coordinates and a time coordinate; the totality of physical "events" is thus thought of as being embedded in a four-dimensional continuous manifold. But on the basis of classical mechanics this four-dimensional continuum breaks up objectively into the one-dimensional time and into three-dimensional spatial sections, only the latter of which contain simultaneous events. This resolution is the same for all inertial systems. The simultaneity of two definite events with reference to one inertial system involves the simultaneity of these events in reference to all inertial systems. This is what is meant when we say that the time of classical mechanics is absolute. According to the special theory of relativity it is otherwise. The sum total of events which are simultaneous with a selected event exist, it is true, in relation to a particular inertial system, but no longer independently of the choice of the inertial system. The four-dimensional continuum is now no longer resolvable objectively into sections, all of which contain

at rest explained the Michelson-Morley experiment by assuming that the length of *only* material bodies (along their velocities) contracts, whereas according to the theory of relativity not only material bodes but *any distance (including that between two points in space) contracts*; also the kinematic relativistic effects are not reciprocal in the aether theory.

simultaneous events; "now" loses for the spatially extended world its objective meaning. It is because of this that space and time must be regarded as a four-dimensional continuum that is objectively unresolvable, if it is desired to express the purport of objective relations without unnecessary conventional arbitrariness.

Since the special theory of relativity revealed the physical equivalence of all inertial systems, it proved the untenability of the hypothesis of an aether at rest. It was therefore necessary to renounce the idea that the electromagnetic field is to be regarded as a state of a material carrier. The field thus becomes an irreducible element of physical description, irreducible in the same sense as the concept of matter in the theory of Newton.

Up to now we have directed our attention to finding in what respect the concepts of space and time were *modified* by the special theory of relativity. Let us now focus our attention on those elements which this theory has taken over from classical mechanics. Here also, natural laws claim validity only when an inertial system is taken as the basis of space-time description. The principle of inertia and the principle of the constancy of the velocity of light are valid only with respect to an *inertial system*. The field-laws also can claim to have a meaning and validity only in regard to inertial systems. Thus, as in classical mechanics, space is here also an independent component in the representation of physical reality. If we imagine matter and field to be removed, inertial-space or, more accurately, this space together with the associated time remains behind. The four-dimensional structure (Minkowski-space) is thought of as being the carrier of matter and of the field. Inertial spaces, with their associated times, are only privileged four-dimensional coordinate systems, that are linked together by the linear Lorentz transformations. Since there exist in this four-dimensional structure no longer any sections which represent "now" objectively, the concepts of happening and becoming are indeed not completely suspended, but yet complicated.[6] It appears therefore more natural to think of physical reality as a four-dimensional existence, instead of, as hitherto, the *evolution* of a three-dimensional existence.

This rigid four-dimensional space of the special theory of relativity is to some extent a four-dimensional analogue of H. A. Lorentz's rigid three-dimensional aether. For this theory also the following statement is valid:

---

[6]EDITOR'S NOTE: It is unfortunate that Einstein did not elaborate on this point. I wonder how Minkowski would comment on this. It seems virtually certain he would point out that, as all events of spacetime exist *equally*, the pre-relativistic concepts of happening and becoming do not represent anything in spacetime because these concepts *presuppose* that only the 'now' events exist (according to the local becoming view, which amounts either to event solipsism or relativization of existence, at least one event should be real as 'now' in spacetime), whereas *the very definition of spacetime is based on the equal existence of all events*; that is why the expression "becoming in spacetime" is a textbook contradiction in terms. And Einstein himself, *de facto*, states the same thing (what Minkowski would say) in the next sentence – that one should "think of physical reality as a four-dimensional existence;" again – such existence, *by its very definition*, means that *all* moments of time (of any observer) are given *en bloc* since they form the fourth dimension and therefore *no* becoming (presupposing that only the present event(s) is / are real) is possible in a four-dimensional existence.

The description of physical states postulates space as being initially given and as existing independently. Thus even this theory does not dispel Descartes' uneasiness concerning the independent, or indeed, the *a priori* existence of "empty space". The real aim of the elementary discussion given here is to show to what extent these doubts are overcome by the general theory of relativity.

The concept of space in the general theory of relativity

This theory arose primarily from the endeavour to understand the equality of inertial and gravitational mass. We start out from an inertial system S_1 whose space is, from the physical point of view, empty. In other words, there exists in the part of space contemplated neither matter (in the usual sense) nor a field (in the sense of the special theory of relativity). With reference to S_1 let there be a second system of reference S_2 in uniform acceleration. Then S_2 is thus not an inertial system. With respect to S_2 every test mass would move with an acceleration, which is independent of its physical and chemical nature. Relative to S_2, therefore, there exists a state which, at least to a first approximation, cannot be distinguished from a gravitational field. The following concept is thus compatible with the observable facts: S_2 is also equivalent to an "inertial system"; but with respect to S_2 a (homogeneous) gravitational field is present (about the origin of which one does not worry in this connection). Thus when the gravitational field is included in the framework of the consideration, the inertial system loses its objective significance, assuming that this "principle of equivalence" can be extended to any relative motion whatsoever of the systems of reference. If it is possible to base a consistent theory on these fundamental ideas, it will satisfy of itself the fact of the equality of inertial and gravitational mass, which is strongly confirmed empirically.

Considered four-dimensionally, a non-linear transformation of the four coordinates corresponds to the transition from S_1 to S_2. The question now arises: What kind of non-linear transformations are to be permitted, or, how is the Lorentz transformation to be generalised? In order to answer this question, the following consideration is decisive.

We ascribe to the inertial system of the earlier theory this property: Differences in coordinates are measured by stationary "rigid" measuring rods, and differences in time by clocks at rest. The first assumption is supplemented by another, namely, that for the relative laying out and fitting together of measuring rods at rest, the theorems on "lengths" in Euclidean geometry hold. From the results of the special theory of relativity it is then concluded, by elementary considerations, that this direct physical interpretation of the coordinates is lost for systems of reference (S_2) accelerated relatively to inertial systems (S_1). But if this is the case, the coordinates now express only the order or rank of the "contiguity" and hence also the dimensional grade of the space, but do not express any of its metrical properties. We are thus led to extend the transformations to arbitrary continuous transformations.[7] This implies the general principle of relativity: Natural laws must be covariant with respect to arbitrary continuous transformations of the coordinates. This requirement (combined with that of the greatest possible logical simplicity of the laws) limits the natural laws concerned incomparably more strongly than the special principle of relativity.

This train of ideas is based essentially on the field as an independent concept. For the conditions prevailing with respect to S_2 are interpreted as a gravitational field, without the question of the existence of masses which produce this field being raised. By virtue of this train of ideas it can also be grasped why the laws of the pure gravitational field are more directly linked with the idea of general relativity than the laws for fields of a general kind (when, for instance, an electromagnetic field is present). We have, namely, good ground for the assumption that the "field-free", Minkowski-space represents a special case possible in natural law, in fact, the simplest conceivable special case. With respect to its metrical character, such a space is characterised by the fact that $dx_1^2 + dx_2^2 + dx_3^2$ is the square of the spatial separation, measured with a unit gauge, of two infinitesimally neighbouring points of a three-dimensional "space-like" cross section (Pythagorean theorem), whereas dx^4 is the temporal separation, measured with a suitable time gauge, of two events with common (x_1, x_2, x_3). All this simply means that an objective metrical significance is attached to the quantity

$$ds^2 = dx_1^2 + dx_2^2 + dx_3^2 - dx_4^2 \tag{1}$$

as is readily shown with the aid of the Lorentz transformations. Mathematically, this fact corresponds to the condition that ds^2 is invariant with respect to Lorentz transformations. If now, in the sense of the general principle of relativity, this space (cf. eq. (1)) is subjected to an arbitrary continuous transformation of the coordinates, then the objectively significant quantity ds is expressed in the new system of coordinates by the relation

$$ds^2 = g_{ik} dx_i dx_k \ldots \tag{1a}$$

which has to be summed up over the indices i and k for all combinations 11, 12, ... up to 44. The terms g_{ik} now are not constants, but functions of the coordinates, which are determined by the arbitrarily chosen transformation. Nevertheless, the terms g_{ik} are not arbitrary functions of the new coordinates, but just functions of such a kind that the form (1a) can be transformed back again into the form (1) by a continuous transformation of the four coordinates. In order that this may be possible, the functions g_{ik} must satisfy certain general covariant equations of condition, which were derived by B. Riemann more than half a century before the formulation of the general theory of relativity ("Riemann condition"). According to the principle of equivalence, (1a) describes in general covariant form a gravitational field of a special kind, when the functions g_{ik} satisfy the Riemann condition.

---

[7]This inexact mode of expression will perhaps suffice here.

It follows that the law for the pure gravitational field of a general kind must be satisfied when the Riemann condition is satisfied; but it must be weaker or less restricting than the Riemann condition. In this way the field law of pure gravitation is practically completely determined, a result which will not be justified in greater detail here.

We are now in a position to see how far the transition to the general theory of relativity modifies the concept of space. In accordance with classical mechanics and according to the special theory of relativity, space (space-time) has an existence independent of matter or field. In order to be able to describe at all that which fills up space and is dependent on the coordinates, space-time or the inertial system with its metrical properties must be thought of at once as existing, for otherwise the description of "that which fills up space" would have no meaning.[8] On the basis of the general theory of relativity, on the other hand, space as opposed to "what fills space", which is dependent on the coordinates, has no separate existence. Thus a pure gravitational field might have been described in terms of the g_{ik} (as functions of the coordinates), by solution of the gravitational equations. If we imagine the gravitational field, i.e. the functions g_{ik}, to be removed, there does not remain a space of the type (1), but absolutely *nothing*, and also no "topological space". For the functions g_{ik} describe not only the field, but at the same time also the topological and metrical structural properties of the manifold. A space of the type (1), judged from the standpoint of the general theory of relativity, is not a space without field, but a special case of the g_{ik} field, for which—for the coordinate system used, which in itself has no objective significance—the functions g_{ik} have values that do not depend on the coordinates. There is no such thing as an empty space, i.e. a space without field. Space-time does not claim existence on its own, but only as a structural quality of the field.

Thus Descartes was not so far from the truth when he believed he must exclude the existence of an empty space. The notion indeed appears absurd, as long as physical reality is seen exclusively in ponderable bodies. It requires the idea of the field as the representative of reality, in combination with the general principle of relativity, to show the true kernel of Descartes' idea; there exists no space "empty of field".

Generalised theory of gravitation

The theory of the pure gravitational field on the basis of the general theory of relativity is therefore readily obtainable, because we may be confident that the "field-free" Minkowski space with its metric in conformity with (1) must satisfy the general laws of field. From this special case the law of gravitation follows by a generalisation which is practically free from arbitrariness. The further development of the theory is not so unequivocally

---

[8]If we consider that which fills space (e.g. the field) to be removed, there still remains the metric space in accordance with (1), which would also determine the inertial behaviour of a test body introduced into it.

determined by the general principle of relativity; it has been attempted in various directions during the last few decades. It is common to all these attempts, to conceive physical reality as a field, and moreover, one which is a generalisation of the gravitational field, and in which the field law is a generalisation of the law for the pure gravitational field. After long probing I believe that I have now found[9] the most natural form for this generalisation, but I have not yet been able to find out whether this generalised law can stand up against the facts of experience.

The question of the particular field law is secondary in the preceding general considerations. At the present time, the main question is whether a field theory of the kind here contemplated can lead to the goal at all. By this is meant a theory which describes exhaustively physical reality, including four-dimensional space, by a field. The present-day generation of physicists is inclined to answer this question in the negative. In conformity with the present form of the quantum theory, it believes that the state of a system cannot be specified directly, but only in an indirect way by a statement of the statistics of the results of measurement attainable on the system. The conviction prevails that the experimentally assured duality of nature (corpuscular and wave structure) can be realised only by such a weakening of the concept of reality. I think that such a far-reaching theoretical renunciation is not for the present justified by our actual knowledge, and that one should not desist from pursuing to the end the path of the relativistic field theory.

---

[9]The generalisation can be characterised in the following way. In accordance with its derivation from empty "Minkowski space", the pure gravitational field of the functions g_{ik} has the property of symmetry given by $g_{ik} = g_{ki}$ ($g_{12} = g_{21}$, etc.). The generalised field is of the same kind, but without this property of symmetry. The derivation of the field law is completely analogous to that of the special case of pure gravitation.

BIBLIOGRAPHY

BIOGRAPHICAL

Out of My Later Years: Albert Einstein. (Constable, 1950)
Einstein—His Life and Times: Philipp Frank. (Cape, 1948)

INTRODUCTORY OR GENERAL

The Special Theory of Relativity: H. Dingle. (Methuen, 1940)
The Expanding Universe: A. Eddington. (Cambridge, 1933)
Space, Time and Gravitation: A. Eddington. (Cambridge, 1923)
The Meaning of Relativity: A. Einstein. (Fifth revised edition; Methuen, 1951)
Evolution of Physics: A. Einstein and L. Infeld. (Cambridge, 1947)
Relativity Physics: W. H. McCrea. (Methuen, 1947)
Albert Einstein: Philosopher—Scientist. (The Library of Living Philosophers, Vol. VII.) Edited by P. A. Schilpp. (Cambridge, 1950)
Space-Time Structure: E. Schrödinger. (Cambridge, 1950)
The Structure of the Universe: G. J. Whitrow. (Hutchinson's University Library, No. 29, 1949)

MATHEMATICAL

Introduction to the Theory of Relativity: P. G. Bergmann. (Prentice-Hall, 1942)
The Mathematical Theory of Relativity: A. Eddington. (Cambridge, 1924)
Relativity, Gravitation and World Structure: E. A. Milne. (Oxford, 1935)
Kinematic Relativity: E. A. Milne. (Oxford, 1948)
The Theory of Relativity: C. Møller. (Oxford, 1952)
Relativitätstheorie: W. Pauli, Jr. (Sonderabdruck aus der Enzyklopädie der mathematischen Wissenschaften, V (2), 543–773). (Teubner, Leipzig, 1921)
Relativity, Thermodynamics and Cosmology: R. C. Tolman. (Oxford, 1934)
Space, Time and Matter: H. Weyl. (Methuen, 1922)

I am indebted to my former colleagues Dr. B. Donovan (Northern Polytechnic, London) and Professor A. G. Walker (Liverpool University) for suggestions in the choice of books.

R. W. L.

Appendix (by the Editor): On Relativistic Mass

"Mass is a mess" [1]. Not quite...

Summary

The relevant relativistic experimental evidence unambiguously demonstrates that *relativistic mass is an experimental fact*. Like mass, which reflects the experimental fact that a body resists its acceleration (because mass is defined as the measure of that resistance), relativistic mass also reflects an experimental fact – the increasing resistance a body offers when accelerated to velocities close to that of light (and relativistic mass is similarly defined as the measure of that *increasing* resistance as the body's velocity increases).

During the last three decades physicists have witnessed (or rather endured) "what has probably been the most vigorous campaign ever waged against the concept of relativistic mass"[1] [2].

It seems that campaign had been prompted by Adler's paper "Does mass really depend on velocity, dad?" [3] in which he had even discovered support for his denial of relativistic mass in Einstein's view on this concept [3, p. 742]:

> Whatever Einstein's precise early views were on the subject, his view in later life appears clear. In a 1948 letter to Lincoln Barnett, he wrote

> "It is not good to introduce the concept of the mass $M = m/(1 - v^2/c^2)^{1/2}$ of a body for which no clear definition can be given. It is better to introduce no other mass than the 'rest mass' m. Instead of introducing M, it is better to mention the expression for the momentum and energy of a body in motion."

Unfortunately, Einstein's unclear view of the relativistic mass[2] appears to have provided some encouragement for the campaign against the use of relativistic mass, but the above quote does not in fact demonstrate that

---

[1] For a detailed account of the controversy over relativistic mass see Chapter 2 of Max Jammer's excellent book *Concepts of Mass in Contemporary Physics and Philosophy* [2].

[2] In his 1905 paper [4] Einstein defined two relativistic masses – longitudinal and transverse masses – but later avoided the entire concept of relativistic mass, including in this book. See [5], [6].

"his view in later life appears clear" – Einstein merely expresses his concern and reservation about the *definition* of M; this becomes evident when it is taken into account that the translation of the above part of Einstein's letter is inaccurate and misleading – compare the translation by Ruschin (there are no such phrases as "introduce no other mass than" and "Instead of introducing M") [7] :

> The German word *daneben* does not mean "instead of," but rather "besides," "in addition to" or "moreover." I would therefore translate the passage:
>
> It is not proper to speak of the mass $M = m/(1 - v^2/c^2)^{1/2}$ of a moving body, because no clear definition can be given for M. It is preferable to restrict oneself to the "rest mass" m. Besides, one may well use the expression for momentum and energy when referring to the inertial behavior of rapidly moving bodies.[3]

Two years after Adler's paper L. B. Okun started a series of publications [8]-[14], which seem to had been the driving force behind the unprecedented campaign against the concept of relativistic mass. In May 1990 *Physics Today* published a number of letters to the Editor with comments on Okun's first article [8] and Okun's replies. W. Rindler's reaction was the sharpest [15]:

> I am disturbed by the harm that Lev Okun's earnest tirade (June 1989 page 31) against the use of relativistic mass ("It is our duty. . . to stop this process") might do to the teaching of relativity. It might suggest to some who have not thought these matters through that there are unresolved logical difficulties in elementary relativity or that if they use the quantity $m = \gamma\, m_0$ they commit some physical blunder, whereas in fact this entire ado is about terminology.

Unfortunately, after that exchange "Okun's polemic condemnation" [2, p. 53] even escalated – here are just two examples of his choice of words: "The pedagogical virus of relativistic mass" (from the abstract of a paper [11]) and "The Virus of Relativistic Mass in the Year of Physics" (a title of a paper published in the volume [12]). I think it is truly sad that such a prominent particle physicist did not seem to have even attempted to entertain the possibility that he might have been fundamentally wrong.

While I share the feeling behind Rindler's reaction, I tend to disagree that "this entire ado is about terminology". And papers in support of the relativistic mass do show that the controversy implies more than terminology (see, for example, [16],[17]); here is the conclusion of Bickerstaff and Patsakos' paper [17, p. 66]:

> Thus we conclude by noting that in answering the elementary question of why two different masses are allowed in relativity,

---

[3] A scan of Einstein's letter in German is included in Okun's article [8].

one obtains a clearer picture of the subject—a picture that is rooted in mathematics and logic rather than semantics and opinion.

Some physicists have argued that "There is no really good definition of mass" [18]-[21], which might explain the relativistic mass controversy. I tend to disagree with this too. The accepted[4] definition of mass – *the mass of a particle is the measure of the resistance the particle offers to its acceleration* – is both adequate for the concept of mass in relativity and does *indisputably* demonstrate that mass indeed increases with velocity and therefore relativistic mass is an integral part of relativity (complementing proper or rest mass):[5]

- a particle whose velocity increases and approaches the velocity of light *offers an increasing resistance to its acceleration,* that is, *obviously,* its mass (the measure of the resistance the particle offers to its acceleration) increases.[6]

- as in relativity the acceleration of a particle is different in different reference frames, the particle's mass is not an invariant since it is not the same in all frames (only the proper or rest mass, measured in the frame in which the particle is at rest, is an invariant).

These facts make the campaign against the concept of relativistic mass both inexplicable and worrisome. Instead of initiating and stimulating research on the origin of relativistic mass (and on the nature of mass in general) in order to achieve a more profound understanding of this fundamental concept in physics,[7] the relativistic mass is not mentioned at all in

---

[4]Only several examples: "Mass is that property of an object that specifies how much resistance an object exhibits to changes in its velocity" [22]; "mass [is] the resistance of a body to a change of motion" [23]; Mass is "the quantitative or numerical measure of a body's inertia, that is of its resistance to being accelerated" [24]; "We use the term *mass* as a quantitative measure of inertia" [25, p. 9-1]; "Mass...measures how hard we have to push a body to achieve a given acceleration" [26]; "Mass is a quantitative measure of inertia...the greater its mass, the more a body "resists" being accelerated" [27]; "*The qualitative definition of the (inertial) mass of a particle is that it is a numerical measure of the reluctance of the particle to being accelerated*" [28]; "mass is a measure of the inertia of an object" [29]; Mass is defined as the "resistance to acceleration" [30]; "Mass is the measure of the gravitational and inertial properties of matter" [31].

[5]As I think it is exceedingly obvious that there are two masses in relativity (like two times; see below) – rest (or proper) mass and relativistic mass – I do not see any need to comment on the problems (coming from the equivalence of mass and energy) with a *single* concept of mass in relativity ("as an invariant, intrinsic property of an object" [32]).

[6]Arguing, effectively, that not the particle's mass but the particle's inertia increases [33] amounts to a rejection of the accepted definition of mass without a valid reason; inertia is the *phenomenon* of offering resistance to acceleration, whereas mass is the *measure* of that resistance.

[7]More research is needed to address the obvious situation: As the *resistance* of a particle to its acceleration depends on the acceleration's direction (the resistance is greater when the acceleration is along the particle's velocity and is becoming infinite as the particle's velocity is approaching the velocity of light), its mass is rather a tensor, not a

many publications[8] (see, for example, the well-known textbook [35]) or, if it is mentioned, it is done to caution the readers[9], that "Most physicists prefer to consider the mass of a particle as fixed" [25, p. 760], that "Most physicists prefer to keep the concept of mass as an invariant, intrinsic property of an object" [32], that "We choose not to use relativistic mass, because it can be a misleading concept" [36] or to warn them [22, p. 1215]:

Watch Out for "Relativistic Mass"

Some older treatments of relativity maintained the conservation of momentum principle at high speeds by using a model in which a particle's mass increases with speed. You might still encounter this notion of "relativistic mass" in your outside reading, especially in older books. Be aware that this notion is no longer widely accepted; today, mass is considered as *invariant*, independent of speed. The mass of an object in all frames is considered to be the mass as measured by an observer at rest with respect to the object.

But phrases such as *"prefer* to consider," *"prefer* to keep," *"choose* not to use" (and *"can be"*), *"no longer widely accepted"* and even "older treatments" do not belong to the rigorous language of physics. Physics is not fashion where expressions such as *"prefer to"* and *"choose not to use,"* for example, naturally fit. Physics at its best asks and addresses questions such as:

- Why is the velocity of light the greatest velocity, which cannot be reached by a particle possessing rest mass?

- Why does such a particle offer an *increasing resistance*[10] as its velocity

---

scalar. In his 1905 paper [4] Einstein defined the two relativistic masses – longitudinal and transverse masses – but later silently abandoned them. With respect to the relativistic masses (longitudinal and transverse) we may witness a repetition of the story with the cosmological constant – initially Einstein used the cosmological constant in his equation linking matter and energy with the spacetime curvature, but later he called it the "biggest blunder of my life;" now cosmologists reintroduced Einstein's cosmological constant. At present time the relativistic mass (let alone the longitudinal and transverse masses) is so out of fashion that even such a prominent relativist as Wolfgang Rindler had to choose the words "confess" and "heuristic" in his letter to the Editor of *Physics Today* [15]: "I will confess to even occasionally using the heuristic concepts of longitudinal mass $\gamma^3 m_0$ and transverse mass γm_0 to predict how a particle will move in a given field of force."

[8]For a list of published works using relativistic mass see [34]. Here I think it is worth mentioning specifically Feynman: "Mass is found to increase with velocity, but appreciable increases require velocities near that of light" [25].

[9]Some authors prefer to take a neutral position: "The use of relativistic mass has its supporters and detractors, some quite strong in their opinions. We will mostly deal with individual particles, so we will sidestep the controversy and use Eq. (37.27) $[\vec{p} = m\vec{v}\,(1-v^2/c^2)^{-1/2}]$ as the generalized definition of momentum with m as a constant for each particle, independent of its state of motion" [27, p. 1244]

[10]In fact, the profound question of the nature of inertia and mass (i.e., the question of the *origin* of the *resistance* a particle offers to its acceleration) has been an open one since Galileo and Newton [37]. The discovery that mass increases with velocity and the controversy over relativistic mass made the need to try to address this open question more urgent.

increases and approaches the velocity of light? Or, which is the same question, why does the mass of a particle increase as its velocity increases and approaches the velocity of light?

No matter how overconfident some physicists (mostly particle physicists) can be, the existing relativistic experimental evidence and the very definition of mass unambiguously demonstrate that the relativistic increase of mass is an *experimental fact* – it is an experimental fact that a particle offers an *increasing resistance* when its velocity increases and approaches the velocity of light and the measure of that *insreasing* resistance is its increasing mass.

Despite the obvious and overwhelming experimental support for the concept of relativistic mass, I believe it may be helpful to address the two most common objections to this concepts.

1. Some authors point out that $\gamma = 1/\sqrt{1 - \beta^2}$ should not be "attached" to the mass, because it comes from the 4-velocity. That γ comes from the 4-velocity is, of course, correct – γ ensures that the velocity of a particle cannot exceed that of light; in other words, γ ensures that no 4-velocity vector, which is timelike, can become lightlike or spacelike. But that is *kinematics*; it says nothing about *dynamics*, that is, it says nothing about why a particle cannot exceed the velocity of light; what is the mechanism that prevents it from doing so. That mechanism is suggested by Newtonian mechanics, where mass is defined as the measure of the *resistance* a particle offers to its acceleration – when Einstein postulated that the velocity of light c is the greatest velocity a particle (with non-zero rest mass) can achieve, it was almost self-evident to assume that a particle would offer an *increasing resistance* when accelerated to velocities approaching that of light, that is, a particle's mass will increase and will approach infinity when the particle's velocity approaches c. And that was repeatedly experimentally confirmed. However, increased resistance and increased relativistic mass are rather only naming the mechanism that prevents a particle from reaching the velocity of light; the origin and nature of the resistance a particle offers when accelerated (an open question in classical physics) and of the increased resistance a particle offers when accelerated to velocities approaching that of light (an open question in spacetime physics) constitutes one of the deepest open questions in spacetime physics.

2. I will summarize my response [39] to the objections of Taylor and Wheeler [40] against using the concept of relativistic mass. Here are their objections:

> The concept of 'relativistic mass' is subject to misunderstanding [...]. First, it applies the name mass – belonging to the magnitude of a 4-vector – to a very different concept, the time component of a 4-vector. Second, it makes increase of energy of an object with velocity or momentum appear to be connected with some change in internal structure of the object. In reality, the increase of energy with velocity originates not in the object but in the geometric properties of spacetime itself.

It is true that the magnitude of the four-momentum is proportional to the rest mass of a particle:

$$|\vec{p}| = mc \, .$$

The time component of the four-momentum

$$p^0 = \frac{mc}{(1 - v^2/c^2)^{1/2}} = m(v)c$$

is proportional to the relativistic mass $m(v) = m(1-v^2/c^2)^{-1/2}$. So the rest (proper) mass m is indeed proportional to the magnitude of a four-vector and is an invariant, whereas the relativistic mass $m(v)$ is a component of a four-vector.

However, the situation is precisely the same with respect to proper time and coordinate time. The square of the spacetime distance Δs^2 between two events lying on a timelike worldline is equal to the scalar product $\Delta\vec{x}\cdot\Delta\vec{x}$ of the displacement four-vector $\Delta\vec{x}$ connecting the two events. In other words, the magnitude of the displacement vector is equal to the spacetime distance along the timelike worldline:

$$|\Delta\vec{x}| = \Delta s \, .$$

As $\Delta s = c\Delta\tau$, the magnitude of $\Delta\vec{x}$ is proportional to the proper time $\Delta\tau$ between the two events on the timelike worldline that are connected by the displacement vector:

$$|\Delta\vec{x}| = c\Delta\tau \, .$$

Therefore, the magnitude of the four-vector $\Delta\vec{x}$ is proportional to the proper time $\Delta\tau$.

On the other hand, however, coordinate time is the zeroth (time) component $\Delta x^0 = c\Delta t$ of the displacement four-vector $\Delta\vec{x}$.

So, if we cannot talk about relativistic mass, by the same argument we should talk only about proper time, which is an invariant, and deny the name 'time' to the coordinate time; however, it is the coordinate time that changes relativistically – the experimentally tested time dilation involves precisely coordinate time.

Therefore, proper or rest mass (which is an invariant) and relativistic mass (which is frame-dependent) are exactly like proper time (which is an invariant) and relativistic / coordinate time (which is frame-dependent) [and, to some extent, like proper and relativistic length].

As we saw above, this becomes even more evident from the very definition of mass as the measure of the *resistance* a particle offers to its acceleration or, in the framework of relativity, as the measure of the resistance a particle offers when deviated from its geodesic path. That resistance is different in different reference frames with respect to which the particle moves with different velocities. Therefore the particle mass should also differ in different frames.

It should be stressed that the resistance (and therefore the increased resistance and energy) arises *in* the particle (more precisely, in the particle's worldtube); it does not come from the geometric properties of spacetime. It is spacetime that determines the shape of a geodesic worldline (and the shape of a geodesic worldtube in the case of a spatially extended particle), but it is the particle that *resists* when prevented from "following" a geodesic path, i.e., when the particle's worldtube is *deformed.*

We have proof that the resistance does not originate in the geometry of spacetime – a particle whose worldtube is deformed due to its deviation from its geodesic shape offers the *same* resistance in *both* flat and curved spacetime as the equivalence of inertial and passive gravitational masses shows (for more details see [39, Chap. 9]).

9 September 2018   Vesselin Petkov
Montreal   *Minkowski Institute*

References

1. W. T. Padgett, Problems with the Current Definitions of Mass, *Physics Essays* **3**, 178–182 (1990)

2. M. Jammer, *Concepts of Mass in Contemporary Physics and Philosophy* (Princeton University Press, Princeton 2000) p. 51

3. C. G. Adler, Does mass really depend on velocity, dad? *American Journal of Physics* **55**, 739 (1987)

4. A. Einstein, On The Electrodynamics Of Moving Bodies, *Annalen der Physik* **17** (1905): 891-921, in *The Collected Papers of Albert Einstein*, Volume 2: *The Swiss Years: Writings, 1900-1909* (Princeton University Press, Princeton 1989), pp. 140-171, p. 169

5. E. Hecht, Einstein on mass and energy, *American Journal of Physics* **77**, 799 (2009)

6. E. Hecht, Einstein Never Approved of Relativistic Mass, The Physics Teacher 47, 336 (2009)

7. S. Ruschin, Putting to Rest Mass Misconceptions, *Physics Today*, May 1990, page 15

8. L. B. Okun, The Concept of Mass, *Physics Today* **42**, 31 (1989)

9. Л. Б. Окунь, Понятие массы, Успехи физических наук, Июль 1989 г. Том 158, вып. 3, с. 511-530 (L. B. Okun, The Concept of Mass, Usp. Fiz. Nauk. 158, 511 (1989) [Sov. Phys. Usp. 32, 629 (1989)])

10. L. B. Okun, Putting to Rest Mass Misconceptions, *Physics Today* **43**, (1990) pp. 15, 115, 117

11. L. B. Okun, The Concept of Mass in the Einstein Year, 2006, arXiv:hep-ph/0602037

12. L. B. Okun, The Virus of Relativistic Mass in the Year of Physics, in: *Gribov Memorial Volume: Quarks, Hadrons, and Strong Interactions*, Proceedings of the Memorial Workshop Devoted to the 75th Birthday of V. N. Gribov, (World Scientific Publishing, Singapore 2009), pp. 470-473

13. L. B. Okun, Mass versus relativistic and rest masses, *American Journal of Physics* **77**, 430 (2009)

14. L. B. Okun, *Energy and Mass in Relativity Theory*, (World Scientific Publishing, Singapore 2009)

15. W. Rindler, Putting to Rest Mass Misconceptions, *Physics Today*, May 1990, page 13

16. T. R. Sandin, In defense of relativistic mass, *American Journal of Physics* **59**, 1032 (1991)

17. R. P. Bickerstaff and G. Patsakos, Relativistic Generalization of Mass, *European Journal of Physics* **16**, 63–68 (1995)

18. E. Hecht, There is no Really Good Definition of Mass, *The Physics Teacher* **44**, 40 (2006)

19. E. Hecht, On Defining Mass, *The Physics Teacher* **49**, 40 (2011)

20. A. Hobson, The Definition of Mass, *The Physics Teacher* **48**, 4 (2010)

21. R. L. Coelho, On the Definition of Mass in Mechanics: Why is it so Difficult? *The Physics Teacher* **50**, 304 (2012)

22. R. A. Serway, J. W. Jewett, Jr., *Physics for Scientists and Engineers with Modern Physics*, 9th ed. (Brooks/Cole, Cengage Learning, Boston 2014) p. 114

23 W. Benenson, J. W. Harris, H. Stocker, H. Lutz (eds), *Handbook of Physics* (Springer, Heidelberg 2002) p. 37

24. S. P. Parker (ed.), *McGraw-Hill Encyclopedia of Physics*, 2nd ed. (McGraw-Hill, New York 1993) p. 762

25. *The Feynman Lectures on Physics, New Millennium Edition*, Vol I (Basic Books, New York 2010) p. 1-2; see also Sections 15-1 (The principle of relativity) and 16-4 (Relativistic mass)

26. M. W. McCall, *Classical Mechanics: From Newton to Einstein: A Modern Introduction*, 2nd ed. (Wiley, Chichester 2011) p. 5

27. H. D. Young, R. A. Freedman, A. L. Ford, *Sears and Zemansky's University Physics with Modern Physics*, 13th ed. (Pearson Education, San Francisco 2012) p. 113

28. D. Gregory, *Classical Mechanics: An Undergraduate Text* (Cambridge University Press, Cambridge 2006) p. 55

29. D. C. Giancoli, *Physics: Principles with Applications*, 7th ed. (Pearson, New York 2014) p. 78

30. W. M. Haynes, D. R. Lide, T. J. Bruno eds, *CRC Handbook of Chemistry and Physics*, 96th ed. (CRC Press, New York, 2015) p. 2-58

31. R. G. Lerner, G. L. Trigg, *Encyclopedia of Physics*, 2nd ed. (VCH Publishers, New York 1991) p. 703

32. S. T. Thornton, A. Rex, *Modern Physics for Scientists and Engineers*, 4th ed. (Brooks/Cole, Cengage Learning, Boston 2013) p. 61

33. J. Roche, What is mass? *European Journal of Physics* **26** (2005) pp. 1-18

34. G. Oas, On the Use of Relativistic Mass in Various Published Works, 2005, arXiv:physics/0504110 [physics.ed-ph]; see also G. Oas, On the Abuse

and Use of Relativistic Mass, 2005, arXiv:physics/0504110 [physics.ed-ph]

35. J. Walker, D. Halliday, R. Resnick, *Fundamentals of Physics Extended*, 10th ed. (John Wiley, New York 2014)

36. K. S. Krane, *Modern Physics* 3rd ed. (Wiley, Chichester 2012) p. 51

37. Newton explicitly defined inertia as the *resistance* a body offers to a change of its velocity (boldface added to "resisting" – V.P.): *"Inherent force of matter is the power of **resisting** by which every body, as far as it is able, perseveres in its state either of resting or of moving uniformly straight forward"* [38]

38. I. Newton, *The Principia: Mathematical Principles of Natural Philosophy*, A new translation by I. Bernard Cohen and Anne Whitman assisted by Julia Budenz (University of California Press, Berkeley 1999) p. 404

39. V. Petkov, *Relativity and the Nature of Spacetime*, 2nd ed. (Springer, Heidelberg 2009) p. 115

40. E. F. Taylor, J. A. Wheeler: *Spacetime Physics: Introduction to Special Relativity*, 2nd ed. (Freeman, New York 1992) pp. 250-251

WHAT IS THE THEORY OF RELATIVITY

Published in *The London Times*, November 28, 1919, translated by R. W. Lawson

I gladly accede to the request of your colleague to write something for The Times (London) on relativity. After the lamentable breakdown of the old active intercourse between men of learning, I welcome this opportunity of expressing my feelings of joy and gratitude toward the astronomers and physicists of England. It is thoroughly in keeping with the great and proud traditions of scientific work in your country that eminent scientists should have spent much time and trouble, and your scientific institutions have spared no expense, to test the implications of a theory which was perfected and published during the war in the land of your enemies. Even though the investigation of the influence of the gravitational field of the sun on light rays is a purely objective matter, I cannot forbear to express my personal thanks to my English colleagues for their work; for without it I could hardly have lived to see the most important implication of my theory tested.

We can distinguish various kinds of theories in physics. Most of them are constructive. They attempt to build up a picture of the more complex phenomena out of the materials of a relatively simple formal scheme from which they start out. Thus the kinetic theory of gases seeks to reduce mechanical, thermal, and diffusional processes to movements of molecules – i.e., to build them up out of the hypothesis of molecular motion. When we say that we have succeeded in understanding a group of natural processes, we invariably mean that a constructive theory has been found which covers the processes in question.

Along with this most important class of theories there exists a second, which I will call "principle-theories." These employ the analytic, not the synthetic, method. The elements which form their basis and starting-point are not hypothetically constructed but empirically discovered ones, general characteristics of natural processes, principles that give rise to mathematically formulated criteria which the separate processes or the theoretical representations of them have to satisfy. Thus the science of thermodynamics seeks by analytical means to deduce necessary conditions, which separate events have to satisfy, from the universally experienced fact that perpetual motion is impossible.

The advantages of the constructive theory are completeness, adaptability, and clearness, those of the principle theory are logical perfection and security of the foundations.

The theory of relativity belongs to the latter class. In order to grasp its nature, one needs first of all to become acquainted with the principles on which it is based. Before I go into these, however, I must observe that the theory of relativity resembles a building consisting of two separate stories, the special theory and the general theory. The special theory, on which the general theory rests, applies to all physical phenomena with the exception of gravitation; the general theory provides the law of gravitation and its relations to the other forces of nature.

It has of course been known since the days of the ancient Greeks that in order to describe the movement of a body, a second body is needed to which the movement of the first is referred. The movement of a vehicle is considered in reference to the earth's surface; that of a planet to the totality of the visible fixed stars. In physics the body to which events are spatially referred is called the coordinate system. The laws of the mechanics of Galileo and Newton, for instance, can only be formulated with the aid of a coordinate system.

The state of motion of the coordinate system may not, however, be arbitrarily chosen, if the laws of mechanics are to be valid (it must be free from rotation and acceleration). A coordinate system which is admitted in mechanics is called an "inertial system." The state of motion of an inertial system is according to mechanics not one that is determined uniquely by nature. On the contrary, the following definition holds good: a coordinate system that is moved uniformly and in a straight line relative to an inertial system is likewise an inertial system. By the "special principle of relativity" is meant the generalization of this definition to include any natural event whatever: thus, every universal law of nature which is valid in relation to a coordinate system C, must also be valid, as it stands, in relation to a coordinate system C', which is in uniform translatory motion relatively to C.

The second principle, on which the special theory of relativity rests, is the "principle of the constant velocity of light *in vacuo*." This principle asserts that light *in vacuo* always has a definite velocity of propagation (independent of the state of motion of the observer or of the source of the light). The confidence which physicists place in this principle springs from the successes achieved by the electrodynamics of Maxwell and Lorentz.

Both the above-mentioned principles are powerfully supported by experience, but appear not to be logically reconcilable. The special theory of relativity finally succeeded in reconciling them logically by a modification of kinematics – i.e., of the doctrine of the laws relating to space and time (from the point of view of physics). It became clear that to speak of the simultaneity of two events had no meaning except in relation to a given coordinate system, and that the shape of measuring devices and the speed at which clocks move depend on their state of motion with respect to the coordinate system.

But the old physics, including the laws of motion of Galileo and Newton, did not fit in with the suggested relativist kinematics. From the latter, general mathematical conditions issued, to which natural laws had to conform, if the above-mentioned two principles were really to apply. To these, physics had to be adapted. In particular, scientists arrived at a new law of motion for (rapidly moving) mass points, which was admirably confirmed in the case of electrically charged particles. The most important upshot of the special theory of relativity concerned the inert masses of corporeal systems. It turned out that the inertia of a system necessarily depends on its energy-content, and this led straight to the notion that inert mass is simply latent energy. The principle of the conservation of mass lost its independence and became fused with that of the conservation of energy.

The special theory of relativity, which was simply a systematic development of the electrodynamics of Maxwell and Lorentz, pointed beyond itself, however. Should the independence of physical laws of the state of motion of the coordinate system be restricted to the uniform translatory motion of coordinate systems in respect to each other? What has nature to do with our coordinate systems and their state of motion? If it is necessary for the purpose of describing nature, to make use of a coordinate system arbitrarily introduced by us, then the choice of its state of motion ought to be subject to no restriction; the laws ought to be entirely independent of this choice (general principle of relativity).

The establishment of this general principle of relativity is made easier by a fact of experience that has long been known, namely, that the weight and the inertia of a body are controlled by the same constant (equality of inertial and gravitational mass). Imagine a coordinate system which is rotating uniformly with respect to an inertial system in the Newtonian manner. The centrifugal forces which manifest themselves in relation to this system must, according to Newton's teaching, be regarded as effects of inertia. But these centrifugal forces are, exactly like the forces of gravity, proportional to the masses of the bodies. Ought it not to be possible in this case to regard the coordinate system as stationary and the centrifugal forces as gravitational forces? This seems the obvious view, but classical mechanics forbid it.

This hasty consideration suggests that a general theory of relativity must supply the laws of gravitation and the consistent following up of the idea has justified our hopes.

But the path was thornier than one might suppose, because it demanded the abandonment of Euclidean geometry. This is to say, the laws according to which solid bodies may be arranged in space do not completely accord with the spatial laws attributed to bodies by Euclidean geometry. This is what we mean when we talk of the "curvature of space." The fundamental concepts of the "straight line," the "plane," etc., thereby lose their precise significance in physics.

In the general theory of relativity the doctrine of space and time, or kinematics, no longer figures as a fundamental independent of the rest of physics. The geometrical behavior of bodies and the motion of clocks rather depend on gravitational fields, which in their turn are produced by matter.

The new theory of gravitation diverges considerably, as regards principles, from Newton's theory. But its practical results agree so nearly with those of Newton's theory that it is difficult to find criteria for distinguishing them which are accessible to experience. Such have been discovered so far:

1. In the revolution of the ellipses of the planetary orbits round the sun (confirmed in the case of Mercury).

2. In the curving of light rays by the action of gravitational fields (confirmed by the English photographs of eclipses).

3. In a displacement of the spectral lines toward the red end of the spectrum in the case of light transmitted to us from stars of considerable magnitude (unconfirmed so far).[1]

The chief attraction of the theory lies in its logical completeness. If a single one of the conclusions drawn from it proves wrong, it must be given up; to modify it without destroying the whole structure seems to be impossible.

Let no one suppose, however, that the mighty work of Newton can really be superseded by this or any other theory. His great and lucid ideas will retain their unique significance for all time as the foundation of our whole modern conceptual structure in the sphere of natural philosophy.

Note: Some of the statements in your paper concerning my life and person owe their origin to the lively imagination of the writer. Here is yet another application of the principle of relativity for the delectation of the reader: today I am described in Germany as a "German savant," and in England as a "Swiss Jew." Should it ever be my fate to be represented as a *bête noire*, I should, on the contrary, become a "Swiss Jew" for the Germans and a "German savant" for the English.

---

[1] EDITOR'S NOTE: This effect had been repeatedly verified later. There had been some initial observations of the gravitational redshift in spectral lines from starts before 1960. But the first definite confirmation of this prediction of general relativity was done by a terrestrial experiment using the Mössbauer effect by Pound and Rebka in 1960 (R. V. Pound and G. A. Rebka, Jr., "Apparent Weight of Photons" *Phys. Rev. Lett.* **4** (1960), 337-341).

Ether and the Theory of Relativity

An Address delivered on May 5th, 1920 in the University of Leyden
Published in A. Einstein, *Sidelights on Relativity* (Methuen & Co. Ltd., London 1922), translated by G. B. Jeffery and W. Perrett, pp. 1-24

How does it come about that alongside of the idea of ponderable matter, which is derived by abstraction from everyday life, the physicists set the idea of the existence of another kind of matter, the ether? The explanation is probably to be sought in those phenomena which have given rise to the theory of action at a distance, and in the properties of light which have led to the undulatory theory. Let us devote a little while to the consideration of these two subjects.

Outside of physics we know nothing of action at a distance. When we try to connect cause and effect in the experiences which natural objects afford us, it seems at first as if there were no other mutual actions than those of immediate contact, e.g. the communication of motion by impact, push and pull, heating or inducing combustion by means of a flame, etc. It is true that even in everyday experience weight, which is in a sense action at a distance, plays a very important part. But since in daily experience the weight of bodies meets us as something constant, something not linked to any cause which is variable in time or place, we do not in everyday life speculate as to the cause of gravity, and therefore do not become conscious of its character as action at a distance. It was Newton's theory of gravitation that first assigned a cause for gravity by interpreting it as action at a distance, proceeding from masses. Newton's theory is probably the greatest stride ever made in the effort towards the causal nexus of natural phenomena. And yet this theory evoked a lively sense of discomfort among Newton's contemporaries, because it seemed to be in conflict with the principle springing from the rest of experience, that there can be reciprocal action only through contact, and not through immediate action at a distance.

It is only with reluctance that man's desire for knowledge endures a dualism of this kind. How was unity to be preserved in his comprehension of the forces of nature? Either by trying to look upon contact forces as being themselves distant forces which admittedly are observable only at a very small distance—and this was the road which Newton's followers, who were entirely under the spell of his doctrine, mostly preferred to take; or by assuming that the Newtonian action at a distance is only *apparently* immediate action at a distance, but in truth is conveyed by a medium permeating space, whether by movements or by elastic deformation of this medium. Thus the endeavour toward a unified view of the nature of forces leads to the hypothesis of an ether. This hypothesis, to be sure, did not at first bring with it any advance in the theory of gravitation or in physics generally, so that it became customary to treat Newton's law of force as an axiom not further reducible. But the ether hypothesis was bound always to play some part in physical science, even if at first only a latent part.

When in the first half of the nineteenth century the far-reaching similarity was revealed which subsists between the properties of light and those of elastic waves in ponderable bodies, the ether hypothesis found fresh support. It appeared beyond question that light must be interpreted as a vibratory process in an elastic, inert medium filling up universal space. It also seemed to be a necessary consequence of the fact that light is capable of polarisation that this medium, the ether, must be of the nature of a solid body, because transverse waves are not possible in a fluid, but only in a solid. Thus the physicists were bound to arrive at the theory of the "quasi-rigid" luminiferous ether, the parts of which can carry out no movements relatively to one another except the small movements of deformation which correspond to light-waves.

This theory—also called the theory of the stationary luminiferous ether—moreover found a strong support in an experiment which is also of fundamental importance in the special theory of relativity, the experiment of Fizeau, from which one was obliged to infer that the luminiferous ether does not take part in the movements of bodies. The phenomenon of aberration also favoured the theory of the quasi-rigid ether.

The development of the theory of electricity along the path opened up by Maxwell and Lorentz gave the development of our ideas concerning the ether quite a peculiar and unexpected turn. For Maxwell himself the ether indeed still had properties which were purely mechanical, although of a much more complicated kind than the mechanical properties of tangible solid bodies. But neither Maxwell nor his followers succeeded in elaborating a mechanical model for the ether which might furnish a satisfactory mechanical interpretation of Maxwell's laws of the electro-magnetic field. The laws were clear and simple, the mechanical interpretations clumsy and contradictory. Almost imperceptibly the theoretical physicists adapted themselves to a situation which, from the standpoint of their mechanical programme, was very depressing. They were particularly influenced by the electro-dynamical investigations of Heinrich Hertz. For whereas they previously had required of a conclusive theory that it should content itself with the fundamental concepts which belong exclusively to mechanics (e.g. densities, velocities, deformations, stresses) they gradually accustomed themselves to admitting electric and magnetic force as fundamental concepts side by side with those of mechanics, without requiring a mechanical interpretation for them. Thus the purely mechanical view of nature was gradually abandoned. But this change led to a fundamental dualism which in the long-run was insupportable. A way of escape was now sought in the reverse direction, by reducing the principles of mechanics to those of electricity, and this especially as confidence in the strict validity of the equations of Newton's mechanics was shaken by the experiments with β-rays and rapid kathode rays.

This dualism still confronts us in unextenuated form in the theory of Hertz, where matter appears not only as the bearer of velocities, kinetic energy, and mechanical pressures, but also as the bearer of electromagnetic fields. Since such fields also occur *in vacuo*—i.e. in free ether—the ether also appears as bearer of electromagnetic fields. The ether appears indistinguishable in its functions from ordinary matter. Within matter it takes part in the motion of matter and in empty space it has everywhere a velocity; so that the ether has a definitely assigned velocity throughout the whole of space. There is no fundamental difference between Hertz's ether and ponderable matter (which in part subsists in the ether).

The Hertz theory suffered not only from the defect of ascribing to matter and ether, on the one hand mechanical states, and on the other hand electrical states, which do not stand in any conceivable relation to each other; it was also at variance with the result of Fizeau's important experiment on the velocity of the propagation of light in moving fluids, and with other established experimental results.

Such was the state of things when H. A. Lorentz entered upon the scene. He brought theory into harmony with experience by means of a wonderful simplification of theoretical principles. He achieved this, the most important advance in the theory of electricity since Maxwell, by taking from ether its mechanical, and from matter its electromagnetic qualities. As in empty space, so too in the interior of material bodies, the ether, and not matter viewed atomistically, was exclusively the seat of electromagnetic fields. According to Lorentz the elementary particles of matter alone are capable of carrying out movements; their electromagnetic activity is entirely confined to the carrying of electric charges. Thus Lorentz succeeded in reducing all electromagnetic happenings to Maxwell's equations for free space.

As to the mechanical nature of the Lorentzian ether, it may be said of it, in a somewhat playful spirit, that immobility is the only mechanical property of which it has not been deprived by H. A. Lorentz. It may be added that the whole change in the conception of the ether which the special theory of relativity brought about, consisted in taking away from the ether its last mechanical quality, namely, its immobility. How this is to be understood will forthwith be expounded.

The space-time theory and the kinematics of the special theory of relativity were modelled on the Maxwell-Lorentz theory of the electromagnetic field. This theory therefore satisfies the conditions of the special theory of relativity, but when viewed from the latter it acquires a novel aspect. For if K be a system of coordinates relatively to which the Lorentzian ether is at rest, the Maxwell-Lorentz equations are valid primarily with reference to K. But by the special theory of relativity the same equations without any change of meaning also hold in relation to any new system of coordinates K' which is moving in uniform translation relatively to K. Now comes the anxious question:–Why must I in the theory distinguish the K system above all K' systems, which are physically equivalent to it in all respects, by assuming that the ether is at rest relatively to the K system? For the theoretician such an asymmetry in the theoretical structure, with no corresponding asymmetry in the system of experience, is intolerable. If we assume the ether to be at rest relatively to K, but in motion relatively to K', the physical equivalence of K and K' seems to me from the logical standpoint, not indeed downright incorrect, but nevertheless inacceptable.

The next position which it was possible to take up in face of this state of things appeared to be the following. The ether does not exist at all. The electromagnetic fields are not states of a medium, and are not bound down to any bearer, but they are independent realities which are not reducible to anything else, exactly like the atoms of ponderable matter. This conception suggests itself the more readily as, according to Lorentz's theory, electromagnetic radiation, like ponderable matter, brings impulse and energy with it, and as, according to the special theory of relativity, both matter and radiation are but special forms of distributed energy, ponderable mass losing its isolation and appearing as a special form of energy.

More careful reflection teaches us, however, that the special theory of relativity does not compel us to deny ether. We may assume the existence of an ether; only we must give up ascribing a definite state of motion to it, i.e. we must by abstraction take from it the last mechanical characteristic which Lorentz had still left it. We shall see later that this point of view, the conceivability of which I shall at once endeavour to make more intelligible by a somewhat halting comparison, is justified by the results of the general theory of relativity.

Think of waves on the surface of water. Here we can describe two entirely different things. Either we may observe how the undulatory surface forming the boundary between water and air alters in the course of time; or else—with the help of small floats, for instance—we can observe how the position of the separate particles of water alters in the course of time. If the existence of such floats for tracking the motion of the particles of a fluid were a fundamental impossibility in physics—if, in fact, nothing else whatever were observable than the shape of the space occupied by the water as it varies in time, we should have no ground for the assumption that water consists of movable particles. But all the same we could characterise it as a medium.

We have something like this in the electromagnetic field. For we may picture the field to ourselves as consisting of lines of force. If we wish to interpret these lines of force to ourselves as something material in the ordinary sense, we are tempted to interpret the dynamic processes as motions of these lines of force, such that each separate line of force is tracked through the course of time. It is well known, however, that this way of regarding the electromagnetic field leads to contradictions.

Generalising we must say this:—There may be supposed to be extended physical objects to which the idea of motion cannot be applied. They may not be thought of as consisting of particles which allow themselves to be separately tracked through time. In Minkowski's idiom this is expressed as follows:—Not every extended conformation in the four-dimensional world can be regarded as composed of world-threads. The special theory of relativity forbids us to assume the ether to consist of particles observable through time, but the hypothesis of ether in itself is not in conflict with the special theory of relativity. Only we must be on our guard against ascribing a state of motion to the ether.

Certainly, from the standpoint of the special theory of relativity, the ether hypothesis appears at first to be an empty hypothesis. In the equations of the electromagnetic field there occur, in addition to the densities of the electric charge, *only* the intensities of the field. The career of electromagnetic processes *in vacuo* appears to be completely determined by these equations, uninfluenced by other physical quantities. The electromagnetic fields appear as ultimate, irreducible realities, and at first it seems superfluous to postulate a homogeneous, isotropic ether-medium, and to envisage electromagnetic fields as states of this medium.

But on the other hand there is a weighty argument to be adduced in favour of the ether hypothesis. To deny the ether is ultimately to assume that empty space has no physical qualities whatever. The fundamental facts of mechanics do not harmonize with this view. For the mechanical behaviour of a corporeal system hovering freely in empty space depends not only on relative positions (distances) and relative velocities, but also on its state of rotation, which physically may be taken as a characteristic not appertaining to the system in itself. In order to be able to look upon the rotation of the system, at least formally, as something real, Newton objectivises space. Since he classes his absolute space together with real things, for him rotation relative to an absolute space is also something real. Newton might no less well have called his absolute space "Ether"; what is essential is merely that besides observable objects, another thing, which is not perceptible, must be looked upon as real, to enable acceleration or rotation to be looked upon as something real.

It is true that Mach tried to avoid having to accept as real something which is not observable by endeavouring to substitute in mechanics a mean acceleration with reference to the totality of the masses in the universe in place of an acceleration with reference to absolute space. But inertial resistance opposed to relative acceleration of distant masses presupposes action at a distance; and as the modern physicist does not believe that he may accept this action at a distance, he comes back once more, if he follows Mach, to the ether, which has to serve as medium for the effects of inertia. But this conception of the ether to which we are led by Mach's way of thinking differs essentially from the ether as conceived by Newton, by Fresnel, and by Lorentz. Mach's ether not only *conditions* the behaviour of inert masses, but is also *conditioned* in its state by them.

Mach's idea finds its full development in the ether of the general theory of relativity. According to this theory the metrical qualities of the continuum of space-time differ in the environment of different points of space-time, and are partly *conditioned* by the matter existing outside of the territory under consideration. This space-time variability of the reciprocal relations of the standards of space and time, or, perhaps, the recognition of the fact that "empty space" in its physical relation is neither homogeneous nor isotropic, compelling us to describe its state by ten functions (the gravitation potentials $g_{\mu\nu}$), has, I think, finally disposed of the view that space is physically empty. But therewith the conception of the ether has again acquired an intelligible content, although this content differs widely from that of the ether of the mechanical undulatory theory of light. The ether of the general theory of relativity is a medium which is itself devoid of *all* mechanical and kinematical qualities, but helps to determine mechanical (and electromagnetic) events.

What is fundamentally new in the ether of the general theory of relativity as opposed to the ether of Lorentz consists in this, that the state of the former is at every place determined by connections with the matter and the state of the ether in neighbouring places, which are amenable to law in the form of differential equations; whereas the state of the Lorentzian ether in the absence of electromagnetic fields is conditioned by nothing outside itself, and is everywhere the same. The ether of the general theory of relativity is transmuted conceptually into the ether of Lorentz if we substitute constants for the functions of space which describe the former, disregarding the causes which condition its state. Thus we may also say, I think, that the ether of the general theory of relativity is the outcome of the Lorentzian ether, through relativation.

As to the part which the new ether is to play in the physics of the future we are not yet clear. We know that it determines the metrical relations in the space-time continuum, e.g. the configurative possibilities of solid bodies as well as the gravitational fields; but we do not know whether it has an essential share in the structure of the electrical elementary particles constituting matter. Nor do we know whether it is only in the proximity of ponderable masses that its structure differs essentially from that of the Lorentzian ether; whether the geometry of spaces of cosmic extent is approximately Euclidean. But we can assert by reason of the relativistic equations of gravitation that there must be a departure from Euclidean relations, with spaces of cosmic order of magnitude, if there exists a positive mean density, no matter how small, of the matter in the universe. In this case the universe must of necessity be spatially unbounded and of finite magnitude, its magnitude being determined by the value of that mean density.

If we consider the gravitational field and the electromagnetic field from the stand-point of the ether hypothesis, we find a remarkable difference between the two. There can be no space nor any part of space without gravitational potentials; for these confer upon space its metrical qualities, without which it cannot be imagined at all. The existence of the gravitational field is inseparably bound up with the existence of space. On the other hand a part of space may very well be imagined without an electromagnetic field; thus in contrast with the gravitational field, the electromagnetic field seems to be only secondarily linked to the ether, the formal nature of the electromagnetic field being as yet in no way determined by that of gravitational ether. From the present state of theory it looks as if the electromagnetic field, as opposed to the gravitational field, rests upon an entirely new formal *motif*, as though nature might just as well have endowed the gravitational ether with fields of quite another type, for example, with fields of a scalar potential, instead of fields of the electromagnetic type.

Since according to our present conceptions the elementary particles of matter are also, in their essence, nothing else than condensations of the electromagnetic field, our present view of the universe presents two realities which are completely separated from each other conceptually, although connected causally, namely, gravitational ether and electromagnetic field, or—as they might also be called—space and matter.

Of course it would be a great advance if we could succeed in comprehending the gravitational field and the electromagnetic field together as one unified conformation. Then for the first time the epoch of theoretical physics founded by Faraday and Maxwell would reach a satisfactory conclusion. The contrast between ether and matter would fade away, and, through the general theory of relativity, the whole of physics would become a complete system of thought, like geometry, kinematics, and the theory of gravitation. An exceedingly ingenious attempt in this direction has been made by the mathematician H. Weyl; but I do not believe that his theory will hold its ground in relation to reality. Further, in contemplating the immediate future of theoretical physics we ought not unconditionally to reject the possibility that the facts comprised in the quantum theory may set bounds to the field theory beyond which it cannot pass.

Recapitulating, we may say that according to the general theory of relativity space is endowed with physical qualities; in this sense, therefore, there exists an ether. According to the general theory of relativity space without ether is unthinkable; for in such space there not only would be no propagation of light, but also no possibility of existence for standards of space and time (measuring-rods and clocks), nor therefore any space-time intervals in the physical sense. But this ether may not be thought of as endowed with the quality characteristic of ponderable media, as consisting of parts which may be tracked through time. The idea of motion may not be applied to it.

Geometry and Experience

An expanded form of an Address to the Prussian Academy of Sciences in Berlin on January 27th, 1921

Published in A. Einstein, *Sidelights on Relativity* (Methuen & Co. Ltd., London 1922), translated by G. B. Jeffery and W. Perrett, pp. 25-56

One reason why mathematics enjoys special esteem, above all other sciences, is that its laws are absolutely certain and indisputable, while those of all other sciences are to some extent debatable and in constant danger of being overthrown by newly discovered facts. In spite of this, the investigator in another department of science would not need to envy the mathematician if the laws of mathematics referred to objects of our mere imagination, and not to objects of reality. For it cannot occasion surprise that different persons should arrive at the same logical conclusions when they have already agreed upon the fundamental laws (axioms), as well as the methods by which other laws are to be deduced therefrom. But there is another reason for the high repute of mathematics, in that it is mathematics which affords the exact natural sciences a certain measure of security, to which without mathematics they could not attain.

At this point an enigma presents itself which in all ages has agitated inquiring minds. How can it be that mathematics, being after all a product of human thought which is independent of experience, is so admirably appropriate to the objects of reality? Is human reason, then, without experience, merely by taking thought, able to fathom the properties of real things.

In my opinion the answer to this question is, briefly, this:—As far as the laws of mathematics refer to reality, they are not certain; and as far as they are certain, they do not refer to reality. It seems to me that complete clearness as to this state of things first became common property through that new departure in mathematics which is known by the name of mathematical logic or "Axiomatics." The progress achieved by axiomatics consists in its having neatly separated the logical-formal from its objective or intuitive content; according to axiomatics the logical-formal alone forms the subject-matter of mathematics, which is not concerned with the intuitive or other content associated with the logical-formal.

Let us for a moment consider from this point of view any axiom of geometry, for instance, the following:—Through two points in space there always passes one and only one straight line. How is this axiom to be interpreted in the older sense and in the more modern sense?

The older interpretation:—Every one knows what a straight line is, and what a point is. Whether this knowledge springs from an ability of the human mind or from experience, from some collaboration of the two or from some other source, is not for the mathematician to decide. He leaves the question to the philosopher. Being based upon this knowledge, which precedes all mathematics, the axiom stated above is, like all other axioms, self-evident, that is, it is the expression of a part of this *à priori* knowledge.

The more modern interpretation:—Geometry treats of entities which are denoted by the words straight line, point, etc. These entities do not take for granted any knowledge or intuition whatever, but they presuppose only the validity of the axioms, such as the one stated above, which are to be taken in a purely formal sense, i.e. as void of all content of intuition or experience. These axioms are free creations of the human mind. All other propositions of geometry are logical inferences from the axioms (which are to be taken in the nominalistic sense only). The matter of which geometry treats is first defined by the axioms. Schlick in his book on epistemology has therefore characterised axioms very aptly as "implicit definitions."

This view of axioms, advocated by modern axiomatics, purges mathematics of all extraneous elements, and thus dispels the mystic obscurity which formerly surrounded the principles of mathematics.

But a presentation of its principles thus clarified makes it also evident that mathematics as such cannot predicate anything about perceptual objects or real objects. In axiomatic geometry the words "point," "straight line," etc., stand only for empty conceptual schemata. That which gives them substance is not relevant to mathematics.

Yet on the other hand it is certain that mathematics generally, and particularly geometry, owes its existence to the need which was felt of learning something about the relations of real things to one another. The very word geometry, which, of course, means earth-measuring, proves this. For earth-measuring has to do with the possibilities of the disposition of certain natural objects with respect to one another, namely, with parts of the earth, measuring-lines, measuring-wands, etc. It is clear that the system of concepts of axiomatic geometry alone cannot make any assertions as to the relations of real objects of this kind, which we will call practically-rigid bodies. To be able to make such assertions, geometry must be stripped of its merely logical-formal character by the co-ordination of real objects of experience with the empty conceptual frame-work of axiomatic geometry. To accomplish this, we need only add the proposition:—Solid bodies are related, with respect to their possible dispositions, as are bodies in Euclidean geometry of three dimensions. Then the propositions of Euclid contain affirmations as to the relations of practically-rigid bodies.

Geometry thus completed is evidently a natural science; we may in fact regard it as the most ancient branch of physics. Its affirmations rest essentially on induction from experience, but not on logical inferences only. We will call this completed geometry "practical geometry," and shall distinguish it in what follows from "purely axiomatic geometry." The question whether the practical geometry of the universe is Euclidean or not has a clear meaning, and its answer can only be furnished by experience. All linear measurement in physics is practical geometry in this sense, so too is geodetic and astronomical linear measurement, if we call to our help the law of experience that light is propagated in a straight line, and indeed in a straight line in the sense of practical geometry.

I attach special importance to the view of geometry which I have just set forth, because without it I should have been unable to formulate the theory of relativity. Without it the following reflection would have been impossible:—In a system of reference rotating relatively to an inert system, the laws of disposition of rigid bodies do not correspond to the rules of Euclidean geometry on account of the Lorentz contraction; thus if we admit non-inert systems we must abandon Euclidean geometry. The decisive step in the transition to general co-variant equations would certainly not have been taken if the above interpretation had not served as a stepping-stone. If we deny the relation between the body of axiomatic Euclidean geometry and the practically-rigid body of reality, we readily arrive at the following view, which was entertained by that acute and profound thinker, H. Poincaré:—Euclidean geometry is distinguished above all other imaginable axiomatic geometries by its simplicity. Now since axiomatic geometry by itself contains no assertions as to the reality which can be experienced, but can do so only in combination with physical laws, it should be possible and reasonable—whatever may be the nature of reality—to retain Euclidean geometry. For if contradictions between theory and experience manifest themselves, we should rather decide to change physical laws than to change axiomatic Euclidean geometry. If we deny the relation between the practically-rigid body and geometry, we shall indeed not easily free ourselves from the convention that Euclidean geometry is to be retained as the simplest. Why is the equivalence of the practically-rigid body and the body of geometry—which suggests itself so readily—denied by Poincaré and other investigators? Simply because under closer inspection the real solid bodies in nature are not rigid, because their geometrical behaviour, that is, their possibilities of relative disposition, depend upon temperature, external forces, etc. Thus the original, immediate relation between geometry and physical reality appears destroyed, and we feel impelled toward the following more general view, which characterizes Poincaré's standpoint. Geometry (G) predicates nothing about the relations of real things, but only geometry together with the purport (P) of physical laws can do so. Using symbols, we may say that only the sum of (G) + (P) is subject to the control of experience. Thus (G) may be chosen arbitrarily, and also parts of (P); all these laws are conventions. All that is necessary to avoid contradictions is to choose the remainder of (P) so that (G) and the whole of (P) are together in accord with experience. Envisaged in this way, axiomatic geometry and the part of natural law which has been given a conventional status appear as epistemologically equivalent.

Sub specie aeterni Poincaré, in my opinion, is right. The idea of the measuring-rod and the idea of the clock coordinated with it in the theory of relativity do not find their exact correspondence in the real world. It is also clear that the solid body and the clock do not in the conceptual edifice of physics play the part of irreducible elements, but that of composite structures, which may not play any independent part in theoretical physics. But it is my conviction that in the present stage of development of theoretical physics these ideas must still be employed as independent ideas; for we are still far from possessing such certain knowledge of theoretical principles as to be able to give exact theoretical constructions of solid bodies and clocks.

Further, as to the objection that there are no really rigid bodies in nature, and that therefore the properties predicated of rigid bodies do not apply to physical reality,—this objection is by no means so radical as might appear from a hasty examination. For it is not a difficult task to determine the physical state of a measuring-rod so accurately that its behaviour relatively to other measuring-bodies shall be sufficiently free from ambiguity to allow it to be substituted for the "rigid" body. It is to measuring-bodies of this kind that statements as to rigid bodies must be referred.

All practical geometry is based upon a principle which is accessible to experience, and which we will now try to realise. We will call that which is enclosed between two boundaries, marked upon a practically-rigid body, a tract. We imagine two practically-rigid bodies, each with a tract marked out on it. These two tracts are said to be "equal to one another" if the boundaries of the one tract can be brought to coincide permanently with the boundaries of the other. We now assume that:

If two tracts are found to be equal once and anywhere, they are equal always and everywhere.

Not only the practical geometry of Euclid, but also its nearest generalisation, the practical geometry of Riemann, and therewith the general theory of relativity, rest upon this assumption. Of the experimental reasons which warrant this assumption I will mention only one. The phenomenon of the propagation of light in empty space assigns a tract, namely, the appropriate path of light, to each interval of local time, and conversely. Thence it follows that the above assumption for tracts must also hold good for intervals of clock-time in the theory of relativity. Consequently it may be formulated as follows:—If two ideal clocks are going at the same rate at any time and at any place (being then in immediate proximity to each other), they will always go at the same rate, no matter where and when they are again compared with each other at one place.—If this law were not valid for real clocks, the proper frequencies for the separate atoms of the same chemical element would not be in such exact agreement as experience demonstrates. The existence of sharp spectral lines is a convincing experimental proof of the above-mentioned principle of practical geometry. This is the ultimate foundation in fact which enables us to speak with meaning of the mensuration, in Riemann's sense of the word, of the four-dimensional continuum of space-time.

The question whether the structure of this continuum is Euclidean, or in accordance with Riemann's general scheme, or otherwise, is, according to the view which is here being advocated, properly speaking a physical question which must be answered by experience, and not a question of a mere convention to be selected on practical grounds. Riemann's geometry will be the right thing if the laws of disposition of practically-rigid bodies are transformable into those of the bodies of Euclid's geometry with an exactitude which increases in proportion as the dimensions of the part of space-time under consideration are diminished.

It is true that this proposed physical interpretation of geometry breaks down when applied immediately to spaces of sub-molecular order of magnitude. But nevertheless, even in questions as to the constitution of elementary particles, it retains part of its importance. For even when it is a question of describing the electrical elementary particles constituting matter, the attempt may still be made to ascribe physical importance to those ideas of fields which have been physically defined for the purpose of describing the geometrical behaviour of bodies which are large as compared with the molecule. Success alone can decide as to the justification of such an attempt, which postulates physical reality for the fundamental principles of Riemann's geometry outside of the domain of their physical definitions. It might possibly turn out that this extrapolation has no better warrant than the extrapolation of the idea of temperature to parts of a body of molecular order of magnitude.

It appears less problematical to extend the ideas of practical geometry to spaces of cosmic order of magnitude. It might, of course, be objected that a construction composed of solid rods departs more and more from ideal rigidity in proportion as its spatial extent becomes greater. But it will hardly be possible, I think, to assign fundamental significance to this objection. Therefore the question whether the universe is spatially finite or not seems to me decidedly a pregnant question in the sense of practical geometry. I do not even consider it impossible that this question will be answered before long by astronomy. Let us call to mind what the general theory of relativity teaches in this respect. It offers two possibilities:—

1. The universe is spatially infinite. This can be so only if the average spatial density of the matter in universal space, concentrated in the stars, vanishes, i.e. if the ratio of the total mass of the stars to the magnitude of the space through which they are scattered approximates indefinitely to the value zero when the spaces taken into consideration are constantly greater and greater.

2. The universe is spatially finite. This must be so, if there is a mean density of the ponderable matter in universal space differing from zero. The smaller that mean density, the greater is the volume of universal space.

I must not fail to mention that a theoretical argument can be adduced in favour of the hypothesis of a finite universe. The general theory of relativity teaches that the inertia of a given body is greater as there are more ponderable masses in proximity to it; thus it seems very natural to reduce the total effect of inertia of a body to action and reaction between it and the other bodies in the universe, as indeed, ever since Newton's time, gravity has been completely reduced to action and reaction between bodies. From the equations of the general theory of relativity it can be deduced that this total reduction of inertia to reciprocal action between masses—as required by E. Mach, for example—is possible only if the universe is spatially finite.

On many physicists and astronomers this argument makes no impression. Experience alone can finally decide which of the two possibilities is realised in nature. How can experience furnish an answer? At first it might seem possible to determine the mean density of matter by observation of that part of the universe which is accessible to our perception. This hope is illusory. The distribution of the visible stars is extremely irregular, so that we on no account may venture to set down the mean density of star-matter in the universe as equal, let us say, to the mean density in the Milky Way. In any case, however great the space examined may be, we could not feel convinced that there were no more stars beyond that space. So it seems impossible to estimate the mean density. But there is another road, which seems to me more practicable, although it also presents great difficulties. For if we inquire into the deviations shown by the consequences of the general theory of relativity which are accessible to experience, when these are compared with the consequences of the Newtonian theory, we first of all find a deviation which shows itself in close proximity to gravitating mass, and has been confirmed in the case of the planet Mercury. But if the universe is spatially finite there is a second deviation from the Newtonian theory, which, in the language of the Newtonian theory, may be expressed thus:—The gravitational field is in its nature such as if it were produced, not only by the ponderable masses, but also by a mass-density of negative sign, distributed uniformly throughout space. Since this factitious mass-density would have to be enormously small, it could make its presence felt only in gravitating systems of very great extent.

Assuming that we know, let us say, the statistical distribution of the stars in the Milky Way, as well as their masses, then by Newton's law we can calculate the gravitational field and the mean velocities which the stars must have, so that the Milky Way should not collapse under the mutual attraction of its stars, but should maintain its actual extent. Now if the actual velocities of the stars, which can, of course, be measured, were smaller than the calculated velocities, we should have a proof that the actual attractions at great distances are smaller than by Newton's law. From such a deviation it could be proved indirectly that the universe is finite. It would even be possible to estimate its spatial magnitude.

Can we picture to ourselves a three-dimensional universe which is finite, yet unbounded?

The usual answer to this question is "No," but that is not the right answer. The purpose of the following remarks is to show that the answer should be "Yes." I want to show that without any extraordinary difficulty we can illustrate the theory of a finite universe by means of a mental image to which, with some practice, we shall soon grow accustomed.

First of all, an observation of epistemological nature. A geometrical-physical theory as such is incapable of being directly pictured, being merely a system of concepts. But these concepts serve the purpose of bringing a multiplicity of real or imaginary sensory experiences into connection in the mind. To "visualise" a theory, or bring it home to one's mind, therefore means to give a representation to that abundance of experiences for which the theory supplies the schematic arrangement. In the present case we have to ask ourselves how we can represent that relation of solid bodies with respect to their reciprocal disposition (contact) which corresponds to the theory of a finite universe. There is really nothing new in what I have to say about this; but innumerable questions addressed to me prove that the requirements of those who thirst for knowledge of these matters have not yet been completely satisfied.

So, will the initiated please pardon me, if part of what I shall bring forward has long been known?

What do we wish to express when we say that our space is infinite? Nothing more than that we might lay any number whatever of bodies of equal sizes side by side without ever filling space. Suppose that we are provided with a great many wooden cubes all of the same size. In accordance with Euclidean geometry we can place them above, beside, and behind one another so as to fill a part of space of any dimensions; but this construction would never be finished; we could go on adding more and more cubes without ever finding that there was no more room. That is what we wish to express when we say that space is infinite. It would be better to say that space is infinite in relation to practically-rigid bodies, assuming that the laws of disposition for these bodies are given by Euclidean geometry.

Another example of an infinite continuum is the plane. On a plane surface we may lay squares of cardboard so that each side of any square has the side of another square adjacent to it. The construction is never finished; we can always go on laying squares—if their laws of disposition correspond to those of plane figures of Euclidean geometry. The plane is therefore infinite in relation to the cardboard squares. Accordingly we say that the plane is an infinite continuum of two dimensions, and space an infinite continuum of three dimensions. What is here meant by the number of dimensions, I think I may assume to be known.

Now we take an example of a two-dimensional continuum which is finite, but unbounded. We imagine the surface of a large globe and a quantity of small paper discs, all of the same size. We place one of the discs anywhere on the surface of the globe. If we move the disc about, anywhere we like, on the surface of the globe, we do not come upon a limit or boundary anywhere on the journey. Therefore we say that the spherical surface of the globe is an unbounded continuum. Moreover, the spherical surface is a finite continuum. For if we stick the paper discs on the globe, so that no disc overlaps another, the surface of the globe will finally become so full that there is no room for another disc. This simply means that the spherical surface of the globe is finite in relation to the paper discs. Further, the spherical surface is a non-Euclidean continuum of two dimensions, that is to say, the laws of disposition for the rigid figures lying in it do not agree with those of the Euclidean plane. This can be shown in the following way. Place a paper disc on the spherical surface, and around it in a circle place six more discs, each of which is to be surrounded in turn by six discs, and so on. If this construction is made on a plane surface, we have an uninterrupted disposition in which there are six discs touching every disc except those which lie on the outside.

Fig. 1

On the spherical surface the construction also seems to promise success at the outset, and the smaller the radius of the disc in proportion to that of the sphere, the more promising it seems. But as the construction progresses it becomes more and more patent that the disposition of the discs in the manner indicated, without interruption, is not possible, as it should be possible by Euclidean geometry of the the plane surface. In this way creatures which cannot leave the spherical surface, and cannot even peep out from the spherical surface into three-dimensional space, might discover, merely by experimenting with discs, that their two-dimensional "space" is not Euclidean, but spherical space.

From the latest results of the theory of relativity it is probable that our three-dimensional space is also approximately spherical, that is, that the laws of disposition of rigid bodies in it are not given by Euclidean geometry, but approximately by spherical geometry, if only we consider parts of space which are sufficiently great. Now this is the place where the reader's imagination boggles. "Nobody can imagine this thing," he cries indignantly. "It can be said, but cannot be thought. I can represent to myself a spherical surface well enough, but nothing analogous to it in three dimensions."

We must try to surmount this barrier in the mind, and the patient reader will see that it is by no means a particularly difficult task. For this purpose we will first give our attention once more to the geometry of two-dimensional spherical surfaces. In the adjoining figure let K be the spherical surface, touched at S by a plane, E, which, for facility of presentation, is shown in the drawing as a bounded surface. Let L be a disc on the spherical surface. Now let us imagine that at the point N of the spherical surface, diametrically opposite to S, there is a luminous point, throwing a shadow L' of the disc L upon the plane E. Every point on the sphere has its shadow on the plane. If the disc on the sphere K is moved, its shadow

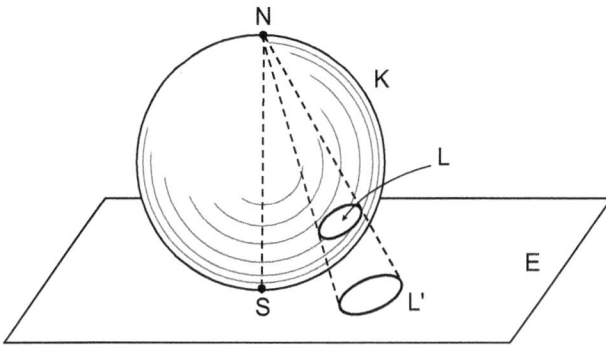

Fig. 2

L' on the plane E also moves. When the disc L is at S, it almost exactly coincides with its shadow. If it moves on the spherical surface away from S upwards, the disc shadow L' on the plane also moves away from S on the plane outwards, growing bigger and bigger. As the disc L approaches the luminous point N, the shadow moves off to infinity, and becomes infinitely great.

Now we put the question, What are the laws of disposition of the disc-shadows L' on the plane E? Evidently they are exactly the same as the laws of disposition of the discs L on the spherical surface. For to each original figure on K there is a corresponding shadow figure on E. If two discs on K are touching, their shadows on E also touch. The shadow-geometry on the plane agrees with the the disc-geometry on the sphere. If we call the disc-shadows rigid figures, then spherical geometry holds good on the plane E with respect to these rigid figures. Moreover, the plane is finite with respect to the disc-shadows, since only a finite number of the shadows can find room on the plane.

At this point somebody will say, "That is nonsense. The disc-shadows are *not* rigid figures. We have only to move a two-foot rule about on the plane E to convince ourselves that the shadows constantly increase in size as they move away from S on the plane towards infinity." But what if the two-foot rule were to behave on the plane E in the same way as the disc-shadows L'? It would then be impossible to show that the shadows increase in size as they move away from S; such an assertion would then no longer have any meaning whatever. In fact the only objective assertion that can be made about the disc-shadows is just this, that they are related in exactly the same way as are the rigid discs on the spherical surface in the sense of Euclidean geometry.

We must carefully bear in mind that our statement as to the growth of the disc-shadows, as they move away from S towards infinity, has in itself no objective meaning, as long as we are unable to employ Euclidean rigid bodies which can be moved about on the plane E for the purpose of comparing the size of the disc-shadows. In respect of the laws of disposition of the shadows L', the point S has no special privileges on the plane any more than on the spherical surface.

The representation given above of spherical geometry on the plane is important for us, because it readily allows itself to be transferred to the three-dimensional case.

Let us imagine a point S of our space, and a great number of small spheres, L', which can all be brought to coincide with one another. But these spheres are not to be rigid in the sense of Euclidean geometry; their radius is to increase (in the sense of Euclidean geometry) when they are moved away from S towards infinity, and this increase is to take place in exact accordance with the same law as applies to the increase of the radii of the disc-shadows L' on the plane.

After having gained a vivid mental image of the geometrical behaviour of our L' spheres, let us assume that in our space there are no rigid bodies at all in the sense of Euclidean geometry, but only bodies having the behaviour of our L' spheres. Then we shall have a vivid representation of three-dimensional spherical space, or, rather of three-dimensional spherical geometry. Here our spheres must be called "rigid" spheres. Their increase in size as they depart from S is not to be detected by measuring with measuring-rods, any more than in the case of the disc-shadows on E, because the standards of measurement behave in the same way as the spheres. Space is homogeneous, that is to say, the same spherical configurations are possible in the environment of all points.[1] Our space is finite, because, in consequence of the "growth" of the spheres, only a finite number of them can find room in space.

In this way, by using as stepping-stones the practice in thinking and visualisation which Euclidean geometry gives us, we have acquired a mental picture of spherical geometry. We may without difficulty impart more depth and vigour to these ideas by carrying out special imaginary constructions. Nor would it be difficult to represent the case of what is called elliptical geometry in an analogous manner. My only aim today has been to show that the human faculty of visualisation is by no means bound to capitulate to non-Euclidean geometry.

---

[1] This is intelligible without calculation—but only for the two-dimensional case—if we revert once more to the case of the disc on the surface of the sphere.

A Brief Outline of the Development of the Theory of Relativity

Published in *Nature*, **106** (No. 2677); February 17, 1921; pp. 782-784, translated by R. W. Lawson

There is something attractive in presenting the evolution of a sequence of ideas in as brief a form as possible, and yet with a completeness sufficient to preserve throughout the continuity of development. We shall endeavour to do this for the Theory of Relativity, and to show that the whole ascent is composed of small, almost self-evident steps of thought.

The entire development starts off from, and is dominated by, the idea of Faraday and Maxwell, according to which all physical processes involve a continuity of action (as opposed to action at a distance), or, in the language of mathematics, they are expressed by partial differential equations. Maxwell succeeded in doing this for electro-magnetic processes in bodies at rest by means of the conception of the magnetic effect of the vacuum-displacement-current, together with the postulate of the identity of the nature of electro-dynamic fields produced by induction, and the electro-static field.

The extension of electro-dynamics to the case of moving bodies fell to the lot of Maxwell's successors. H. Hertz attempted to solve the problem by ascribing to empty space (the aether) quite similar physical properties to those possessed by ponderable matter; in particular, like ponderable matter, the aether ought to have at every point a definite velocity. As in bodies at rest, electromagnetic or magneto-electric induction ought to be determined by the rate of change of the electric or magnetic flow respectively, provided that these velocities of alteration are referred to surface elements moving with the body. But the theory of Hertz was opposed to the fundamental experiment of Fizeau on the propagation of light in flowing liquids. The most obvious extension of Maxwell's theory to the case of moving bodies was incompatible with the results of experiment.

At this point, H. A. Lorentz came to the rescue. In view of his unqualified adherence to the atomic theory of matter, Lorentz felt unable to regard the latter as the seat of continuous electromagnetic fields. He thus conceived of these fields as being conditions of the aether, which was regarded as continuous. Lorentz considered the aether to be intrinsically independent of matter, both from a mechanical and a physical point of view. The aether did not take part in the motions of matter, and a reciprocity between aether and matter could be assumed only in so far as the latter was considered to be the carrier of attached electrical charges. The great value of the theory of Lorentz lay in the fact that the entire electrodynamics of bodies at rest and of bodies in motion was led back to Maxwell's equations of empty space. Not only did this theory surpass that of Hertz from the point of view of method, but with its aid H. A. Lorentz was also pre-eminently successful in explaining the experimental facts.

The theory appeared to be unsatisfactory only in *one* point of fundamental importance. It appeared to give preference to one system of coordinates of a particular state of motion (at rest relative to the aether) as against all other systems of coordinates in motion with respect to this one. In this point the theory seemed to stand in direct opposition to classical mechanics, in which all inertial systems which are in uniform motion with respect to each other are equally justifiable as systems of coordinates (Special Principle of Relativity). In this connection, all experience also in the realm of electro-dynamics (in particular Michelson's experiment) supported the idea of the equivalence of all inertial systems, i.e. was in favour of the special principle of relativity.

The Special Theory of Relativity owes its origin to this difficulty, which, because of its fundamental nature, was felt to be intolerable. This theory originated as the answer to the question: Is the special principle of relativity really contradictory to the field equations of Maxwell for empty space? The answer to this question appeared to be in the affirmative. For if those equations are valid with reference to a system of coordinates K, and we introduce a new system of coordinates K' in conformity with the — to all appearances readily establishable — equations of transformation

$$\left.\begin{array}{l} x' = x - vt \\ y' = y \\ z' = z \\ t' = t \end{array}\right\} \text{ (Galileo transformation),}$$

then Maxwell's field equations are no longer valid in the new coordinates (x', y', z', t'). But appearances are deceptive. A more searching analysis of the physical significance of space and time rendered it evident that the Galileo transformation is founded on arbitrary assumptions, and in particular on the assumption that the statement of simultaneity has a meaning which is independent of the state of motion of the system of coordinates used. It was shown that the field equations for *vacuo* satisfy the special principle of relativity, provided we make use of the equations of transformation stated below:

$$\left.\begin{array}{l} x' = \frac{x - vt}{\sqrt{1 - v^2/c^2}} \\ y' = y \\ z' = z \\ t' = \frac{t - vx/c^2}{\sqrt{1 - v^2/c^2}} \end{array}\right\} \text{ (Lorentz transformation).}$$

In these equations x, y, z represent the coordinates measured with measuring-rods which are at rest with reference to the system of coordinates, and t represents the time measured with suitably adjusted clocks of identical construction, which are in a state of rest.

Now in order that the special principle of relativity may hold, it is necessary that all the equations of physics do not alter their form in the transition from one inertial system to another, when we make use of the Lorentz transformation for the calculation of this change. In the language of mathematics, all systems of equations that express physical laws must be covariant with respect to the Lorentz transformation. Thus, from the point of view of method, the special principle of relativity is comparable to Carnot's principle of the impossibility of perpetual motion of the second kind, for, like the latter, it supplies us with a general condition which all natural laws must satisfy.

Later, H. Minkowski found a particularly elegant and suggestive expression for this condition of covariance, one which reveals a formal relationship between Euclidean geometry of three dimensions and the space-time continuum of physics.

Euclidean Geometry of Three Dimensions

Corresponding to two neighbouring points in space, there exists a numerical measure (distance ds) which conforms to the equation

$$ds^2 = dx_1^2 + dx_2^2 + dx_3^2.$$

It is independent of the system of coordinates chosen, and can be measured with the unit measuring-rod.

The permissible transformations are of such a character that the expression for ds is invariant, i.e. the linear orthogonal transformations are permissible.

With respect to these transformations, the laws of Euclidean geometry are invariant.

Special Theory of Relativity

Corresponding to two neighbouring points in space-time (point events), there exists a numerical measure (distance ds) which conforms to the equation

$$ds^2 = dx_1^2 + dx_2^2 + dx_3^2 + dx_4^2.$$

It is independent of the inertial system chosen, and can be measured with the unit measuring-rod and a standard clock. x_1, x_2, x_3, are here rectangular coordinates, whilst $x_4 = \sqrt{-1}ct$ is the time multiplied by the imaginary unit and by the velocity of light.

The permissible transformations are of such a character that the expression for ds is invariant, i.e. those linear orthogonal substitutions are permissible which maintain the semblance of reality of x_1, x_2, x_3, x_4. These substitutions are the Lorentz transformations.

With respect to these transformations, the laws of physics are invariant.

From this it follows that, in respect of its *role* in the equations of physics, though not with regard to its physical significance, time is equivalent to the space coordinates (apart from the relations of reality). From this point of view, physics is, as it were, a Euclidean geometry of four dimensions, or, more correctly, a statics in a four-dimensional Euclidean continuum.

The development of the special theory of relativity consists of two main steps, namely, the adaptation of the space-time "metrics" to Maxwell's electro-dynamics, and an adaptation of the rest of physics to that altered space-time "metrics." The first of these processes yields the relativity of simultaneity, the influence of motion on measuring-rods and clocks, a modification of kinematics, and in particular a new theorem of addition of velocities. The second process supplies us with a modification of Newton's law of motion for large velocities, together with information of fundamental importance on the nature of inertial mass.

It was found that inertia is not a fundamental property of matter, nor, indeed, an irreducible magnitude, but a property of energy. If an amount of energy E be given to a body, the inertial mass of the body increases by an amount E/c^2, where c is the velocity of light *in vacuo*. On the other hand, a body of mass m is to be regarded as a store of energy of magnitude mc^2.

Furthermore, it was soon found impossible to link up the science of gravitation with the special theory of relativity in a natural manner. In this connection I was struck by the fact that the force of gravitation possesses a fundamental property, which distinguishes it from electro-magnetic forces. All bodies fall in a gravitational field with the same acceleration, or — what is only another formulation of the same fact — the gravitational and inertial masses of a body are numerically equal to each other. This numerical equality suggests identity in character. Can gravitation and inertia be identical? This question leads directly to the General Theory of Relativity. Is it not possible for me to regard the earth as free from rotation, if I conceive of the centrifugal force, which acts on all bodies at rest relatively to the earth, as being a "real" field of gravitation, or part of such a field? If this idea can be carried out, then we shall have proved in very truth the identity of gravitation and inertia. For the same property which is regarded as *inertia* from the point of view of a system not taking part in the rotation can be interpreted as *gravitation* when considered with respect to a system that shares the rotation. According to Newton, this interpretation is impossible, because by Newton's law the centrifugal field cannot be regarded as being produced by matter, and because in Newton's theory there is no place for a "real" field of the "Koriolis-field" type. But perhaps Newton's law of field could be replaced by another that fits in with the field which holds with respect to a "rotating" system of coordinates? My conviction of the identity of inertial and gravitational mass aroused within me the feeling of absolute confidence in the correctness of this interpretation. In this connection I gained encouragement from the following idea. We are familiar with the "apparent" fields which are valid relatively to systems of coordinates possessing arbitrary motion with respect to an inertial system. With the aid of these special fields we should be able to study the law which is satisfied in general by gravitational fields. In this connection we shall have to take account of the fact that the ponderable masses will be the determining factor in producing the field, or, according to the fundamental result of the special theory of relativity, the energy density — a magnitude having the transformational character of a tensor.

On the other hand, considerations based on the metrical results of the special theory of relativity led to the result that Euclidean metrics can no longer be valid with respect to accelerated systems of coordinates. Although it retarded the progress of the theory several years, this enormous difficulty was mitigated by our knowledge that Euclidean metrics holds for small domains. As a consequence, the magnitude ds, which was physically defined in the special theory of relativity hitherto, retained its significance also in the general theory of relativity. But the coordinates themselves lost their direct significance, and degenerated simply into numbers with no physical meaning, the sole purpose of which was the numbering of the space-time points. Thus in the general theory of relativity the coordinates perform the same function as the Gaussian coordinates in the theory of surfaces. A necessary consequence of the preceding is that in such general coordinates the measurable magnitude ds must be capable of representation in the form

$$ds^2 = \sum_{uv} g_{uv} dx_u dx_v,$$

where the symbols g_{uv} are functions of the space-time coordinates. From the above it also follows that the nature of the space-time variation of the factors g_{uv} determines, on one hand the space time metrics, and on the other the gravitational field which governs the mechanical behaviour of material points.

The law of the gravitational field is determined mainly by the following conditions: First, it shall be valid for an arbitrary choice of the system of coordinates; secondly, it shall be determined by the energy tensor of matter; and thirdly, it shall contain no higher differential coefficients of the factors g_{uv} than the second, and must be linear in these. In this way a law was obtained which, although fundamentally different from Newton's law, corresponded so exactly to the latter in the deductions derivable from it that only very few criteria were to be found on which the theory could be decisively tested by experiment.

The following are some of the important questions which are awaiting solution at the present time. Are electrical and gravitational fields really so different in character that there is no formal unit to which they can be reduced? Do gravitational fields play a part in the constitution of matter, and is the continuum within the atomic nucleus to be regarded as appreciably non-Euclidean? A final question has reference to the cosmological problem. Is inertia to be traced to mutual action with distant masses? And connected with the latter: Is the spatial extent of the universe finite? It is here that my opinion differs from that of Eddington. With Mach, I feel that an affirmative answer is imperative, but for the time being nothing can be proved. Not until a dynamical investigation of the large systems of fixed stars has been performed from the point of view of the limits of validity of the Newtonian law of gravitation for immense regions of space will it perhaps be possible to obtain eventually an exact basis for the solution of this fascinating question.

NOTABLE QUOTES BY EINSTEIN FROM THE FIVE WORKS IN THIS VOLUME

On why simultaneity and the one-way velocity of light are conventional (p. 22):

> That light requires the same time to traverse the path $A \longrightarrow M$ as for the path $B \longrightarrow M$ is in reality neither a *supposition nor a hypothesis* about the physical nature of light, but a *stipulation* which I can make of my own freewill in order to arrive at a definition of simultaneity.

Inertial mass is not constant (p. 36):

> [T]he inertial mass of a body is not a constant, but varies according to the change in the energy of the body.

Velocity of light is not constant in general relativity. i.e., in curved space-time:

> p. 58: In the second place our result shows that, according to the general theory of relativity, the law of the constancy of the velocity of light *in vacuo*, which constitutes one of the two fundamental assumptions in the special theory of relativity and to which we have already frequently referred, cannot claim any unlimited validity. A curvature of rays of light can only take place when the velocity of propagation of light varies with position. Now we might think that as a consequence of this, the special theory of relativity and with it the whole theory of relativity would be laid in the dust. But in reality this is not the case. We can only conclude that the special theory of relativity cannot claim an unlimited domain of validity; its results hold only so long as we are able to disregard the influences of gravitational fields on the phenomena (e.g., of light).

> p. 70: In the first part of this book we were able to make use of space-time coordinates which allowed of a simple and direct physical interpretation, and which, according to Chapter 26, can be regarded as four-dimensional Cartesian coordinates. This was possible on the basis of the law of the constancy of

the velocity of light. But according to Chapter 22, the general theory of relativity cannot retain this law. On the contrary, we arrived at the result that according to this latter theory the velocity of light must always depend on the coordinates when a gravitational field is present.

On many spaces (p. 103):

> Before one has become aware of this complication, space appears as an unbounded medium or container in which material objects swim around. But it must now be remembered that there is an infinite number of spaces, which are in motion with respect to each other. The concept of space as something existing objectively and independent of things belongs to prescientific thought, but not so the idea of the existence of an infinite number of spaces in motion relatively to each other. This latter idea is indeed logically unavoidable, but is far from having played a considerable role even in scientific thought.

On why "now" loses its objective meaning and why spacetime is objectively unresolvable (p. 109)

> The four-dimensional continuum is now no longer resolvable objectively into sections, all of which contain simultaneous events; "now" loses for the spatially extended world its objective meaning. It is because of this that space and time must be regarded as a four-dimensional continuum that is objectively unresolvable, if it is desired to express the purport of objective relations without unnecessary conventional arbitrariness.

The above quote in a larger context (p. 108):

> What is the position of the special theory of relativity in regard to the problem of space? In the first place we must guard against the opinion that the four-dimensionality of reality has been newly introduced for the first time by this theory. Even in classical physics the event is localised by four numbers, three spatial coordinates and a time coordinate; the totality of physical "events" is thus thought of as being embedded in a four-dimensional continuous manifold. But on the basis of classical mechanics this four-dimensional continuum breaks up objectively into the one-dimensional time and into three-dimensional spatial sections, only the latter of which contain simultaneous events. This resolution is the same for all inertial systems. The simultaneity of two definite events with reference to one inertial system involves the simultaneity of these events in reference to all inertial systems. This is what is meant when we say that the time of classical mechanics is absolute. According to the

special theory of relativity it is otherwise. The sum total of events which are simultaneous with a selected event exist, it is true, in relation to a particular inertial system, but no longer independently of the choice of the inertial system. The four-dimensional continuum is now no longer resolvable objectively into sections, all of which contain simultaneous events; "now" loses for the spatially extended world its objective meaning. It is because of this that space and time must be regarded as a four-dimensional continuum that is objectively unresolvable, if it is desired to express the purport of objective relations without unnecessary conventional arbitrariness.

On becoming, "now" and four-dimensional existence (p. 109):

Since there exist in this four-dimensional structure no longer any sections which represent "now" objectively, the concepts of happening and becoming are indeed not completely suspended, but yet complicated. It appears therefore more natural to think of physical reality as a four-dimensional existence, instead of, as hitherto, the *evolution* of a three-dimensional existence.

On aether:

p. 109: Since the special theory of relativity revealed the physical equivalence of all inertial systems, it proved the untenability of the hypothesis of an aether at rest. It was therefore necessary to renounce the idea that the electromagnetic field is to be regarded as a state of a material carrier. The field thus becomes an irreducible element of physical description, irreducible in the same sense as the concept of matter in the theory of Newton.

p. 109: This rigid four-dimensional space of the special theory of relativity is to some extent a four-dimensional analogue of H. A. Lorentz's rigid three-dimensional aether.[1]

According to the general theory of relativity space without ether is unthinkable (p. 137):

Recapitulating, we may say that according to the general theory of relativity space is endowed with physical qualities; in this

---

[1] EDITOR'S NOTE: Eddington held the same view – that the aether can be regarded as identical with Minkowski's (four-dimensional) *world*, i.e., spacetime: "apparently there is some divergence of view as to the right name for the fundamental substratum of everything. Since it is the medium the condition of which determined light and electromagnetic force, we may call it *aether*; since it is the subject-matter of the science of geometry, we may call it *space*; sometimes in order to avoid giving preference to either aspect, it is called by Minkowski's term *world*." A. S. Eddington, "Space" or "Aether"? *Nature* **107**, 201 (April 14, 1921); reprinted in A. S. Eddington, *Space, Time and Gravitation: An Outline of the General Relativity Theory* (Minkowski Institute Press, Montreal 2017).

158

sense, therefore, there exists an ether. According to the general theory of relativity space without ether is unthinkable; for in such space there not only would be no propagation of light, but also no possibility of existence for standards of space and time (measuring-rods and clocks), nor therefore any space-time intervals in the physical sense. But this ether may not be thought of as endowed with the quality characteristic of ponderable media, as consisting of parts which may be tracked through time. The idea of motion may not be applied to it.

On the existence of empty spacetime (p. 112):

There is no such thing as an empty space, i.e. a space without field. Space-time does not claim existence on its own, but only as a structural quality of the field.

On the logical completeness of general relativity (p. 130):

The chief attraction of the theory lies in its logical completeness. If a single one of the conclusions drawn from it proves wrong, it must be given up; to modify it without destroying the whole structure seems to be impossible.

No expiration date of Newton's gravitational theory (p. 130):

Let no one suppose, however, that the mighty work of Newton can really be superseded by this or any other theory. His great and lucid ideas will retain their unique significance for all time as the foundation of our whole modern conceptual structure in the sphere of natural philosophy.

On constructive and principle physical theories (p. 158):

We can distinguish various kinds of theories in physics. Most of them are constructive. They attempt to build up a picture of the more complex phenomena out of the materials of a relatively simple formal scheme from which they start out. Thus the kinetic theory of gases seeks to reduce mechanical, thermal, and diffusional processes to movements of molecules – i.e., to build them up out of the hypothesis of molecular motion. When we say that we have succeeded in understanding a group of natural processes, we invariably mean that a constructive theory has been found which covers the processes in question.

Along with this most important class of theories there exists a second, which I will call "principle-theories." These employ the analytic, not the synthetic, method. The elements which form their basis and starting-point are not hypothetically constructed but empirically discovered ones, general characteristics of natural processes, principles that give rise to mathematically formulated criteria which the separate processes or the theoretical representations of them have to satisfy. Thus the science of thermodynamics seeks by analytical means to deduce necessary conditions, which separate events have to satisfy, from the universally experienced fact that perpetual motion is impossible.

The advantages of the constructive theory are completeness, adaptability, and clearness, those of the principle theory are logical perfection and security of the foundations.

The theory of relativity belongs to the latter class. In order to grasp its nature, one needs first of all to become acquainted with the principles on which it is based. Before I go into these, however, I must observe that the theory of relativity resembles a building consisting of two separate stories, the special theory and the general theory. The special theory, on which the general theory rests, applies to all physical phenomena with the exception of gravitation; the general theory provides the law of gravitation and its relations to the other forces of nature.

INDEX

β-rays, 39
"being", 52
"distance", 10, 11
"laws of nature", 55
"mollusk"
    as reference-body, 74
"now", 109, 156, 157
"plain", 9
"point", 9
"straight line", 9
"truth", 9, 17
"world-sphere", 82

aberration, 39, 107, 132
absorption of energy, 36
acceleration, 50, 52, 54, 128
    uniform, 110
acceleration with respect to
    space, 101
accessible to experience, 130
action at a distance, 37
action at a distance), 149
addition of velocities, 18, 31
adjacent points, 66
aether, 40, 107
    at rest, 109, 157
    four-dimensional, 109, 157
    luminiferous
        at rest, 108
    three-dimensional, 109, 157
aether-drift, 40, 41
aether-theory, 107
analytical geometry, 102
ancient Greeks, 128
arbitrary substitutions, 72
astronomical day, 15

astronomy, 12, 75
average density of matter, 99
axiomatic structure, 102
axioms, 9, 93
    truth of, 9

Bachem, 98
basis of theory, 35
becoming, 109, 157
biology, 93
body of reference, 11, 23

Cartesian system of coordinates,
    12, 64, 91
Cartesians, 102
cathode rays, 39
celestial mechanics, 79
centrifugal force, 60, 97
    potential of, 97
centrifugal forces, 129
chest, 52
classical mechanics, 13, 16–18,
    26, 35, 55, 75, 76, 93,
    112, 129
    truth of, 16
clocks, 13, 22, 60, 61, 70, 71,
    73–75, 84, 96, 137, 158
    rate of, 96
concept of acceleration, 101
concept of space, 102
concept of space associated with
    a box, 102
concept of time
    objective, 103
conception of mass, 36
    position, 11

concepts of time and space, 104
conservation of energy, 36, 75
    impulse, 75
    mass, 36, 37
constructive theories, 127, 158
continuity, 71
continuum, 42, 63
    Euclidean, 63, 64, 66, 69
    four-dimensional, 66, 68, 70, 91
    non-Euclidean, 64, 66
    space-time, 59, 68–71
    three-dimensional, 43
    two-dimensional, 70
coordinate
    differences, 68
    differentials, 68
    planes, 27
cosmological problem, 99
cosmological term, 99
Cottingham, 96
counter-point, 83
covariant, 34
Crommelin, 96
curvature of light-rays, 76, 95
    space, 95
curvilinear motion, 57
cyanogen bands, 98

Darwinian theory, 93
Davidson, 96
De Sitter, 19
deductive thought, 93
derivation of laws, 35
Descartes, 101, 110, 112
displacement of spectral lines, 76, 96, 97
distance (line-interval), 10–12, 25, 63, 65, 81
    physical interpretation of, 11
    relativity of, 25
Doppler effect, 107
Doppler principle, 39
Doppler's principle, 99
double stars, 19

eclipse of star, 19

Eddington, 76, 96
electricity, 58
electrodynamics, 16, 20, 33, 35, 58
electrodynamics of Maxwell and Lorentz, 128, 129
electromagnetic field, 109, 157
electromagnetic theory, 39
    waves, 50
electron, 35, 40
    electrical masses of, 40
electrostatics, 58
elements of volume, 105
elliptical space, 83
empirical laws, 93
empty space, 101, 102, 106, 110
encounter (space-time coincidence), 71
equality of inertial and gravitational mass, 110
equivalent, 17
Euclidean geometry, 9, 43, 62, 64, 66, 81, 84, 91, 102, 105, 129
    propositions of, 10, 12
Euclidean space, 43, 64, 91
Evershed, 98
existence
    four-dimensional, 109, 157
    three-dimensional, 109, 157
expanding space, 99
experience, 39, 47

Faraday, 37, 50, 137
Faraday-Maxwell interpretation, 37
Feynman, R., 120
FitzGerald, 41
fixed stars, 15
Fizeau, 32, 39, 40
    experiment of, 32, 149
forces of gravity, 129
four-momentum, 122
four-vector, 122
    displacement, 122
freewill, 22, 155
frequency of atom, 97

Friedmann, 99

galactic line-shift, 99
Galilei, 15
    transformation, 28, 30, 31,
        34, 40
Galileian system of coordinates,
    15–17, 36, 60, 68, 72, 74
Gauss, 64, 65, 67
Gaussian coordinates, 65–67,
    70–74
general principle of relativity,
    111, 129
general relativity
    logical completeness, 130,
        158
general theory of relativity, 47,
    72, 76, 110, 112
geometrical ideas, 9, 10
    propositions, 9
    truth of, 9–10
gravitation, 50, 52, 53, 59, 75
gravitational field, 50, 52, 57, 58,
    70, 72, 74, 84, 110, 127
    potential of, 97
gravitational mass, 51, 53, 75
Grebe, 98
group-density of stars, 79

Helmholtz, 81
Hertz, H., 149
heuristic value of relativity, 34
Hubble, 99
Hume, D., 104

induction, 93
inertia, 51
inertial mass, 51, 53, 74, 75
    not a constant, 36, 155
inertial system, 109, 110, 128
instantaneous photograph
    (snapshot), 88
intensity of gravitational field, 79
intuition, 93
ions, 35

Kepler, 94

kinetic energy, 35, 75
kinetic theory of gases, 127

lattice, 81
law of inertia, 15, 48, 72, 106
laws of Galilei-Newton, 16
    of Nature, 47, 55, 73
laws of gravitation, 129
laws of motion of Galileo and
    Newton, 128
laws of the mechanics of Galileo
    and Newton, 128
Leverrier, 75, 94
light-signal, 28, 87, 89
light-stimulus, 28
limiting velocity (c), 29, 30
lines of force, 79
Lorentz, 107
Lorentz transformation, 28, 31,
    34, 68, 72, 87, 89–91,
    110
    (generalised), 90
    "rotation" in Minkowski's
        "world", 91
Lorentz transformations, 108
Lorentz, H. A., 20, 32, 35, 39–41,
    133, 149
luminiferous ether, 132

Mössbauer effect, 130
Mach, E., 56, 104
magnetic field, 50
mass
    definition, 119
    longitudinal, 37, 117, 120
    passive gravitational, 123
    proper, 119, 122
    relativistic, 117, 119, 121,
        122
    rest, 117, 119, 122
    transverse, 37, 117, 120
mass of heavenly bodies, 98
materialism, 104
matter, 74
Maxwell, 33, 37–39, 107, 137
    fundamental equations, 36,
        58

Maxwell, J.C., 35, 39, 40
Maxwell-Lorentz
    electrodynamics of, 33
Maxwell-Lorentz equations , 40
Maxwell-Lorentz theory, 39, 133
Maxwell's equations, 133
measurement of length, 64
measuring-rod, 11, 25, 61, 75,
    83, 84, 88
measuring-rods, 31, 60, 70, 74,
    137, 158
Mercury, 75, 94
    orbit of, 75, 94
method
    analytic, 127, 158
    synthetic, 127, 158
Michelson, 41
Michelson-Morley experiment,
    41, 107
Minkowski's "world", 91
Minkowski, H., 42–91
Minkowski-space, 109, 111
Morley, 41
motion, 17, 47
    of heavenly bodies, 16, 17,
        35, 75, 84

natural laws must be covariant,
    111
Newcomb, 94
Newton, 15, 56, 75, 79, 94
Newton's
    law of gravitation, 60, 79, 93
    law of motion, 51
Newton's constant of
    gravitation, 97
Newton's law of gravitation, 37
Newtonian mechanics, 105
Newtonian physics, 101
non-Euclidean geometry, 81
non-Galileian reference-bodies,
    72
non-linear transformations, 110
non-uniform motion, 48

object
    "real", 9
    exact, 9
objectivity of space, 102
Okun, L. B., 118
one-way velocity of light
    conventional, 22, 155
    logical circle, 22
optics, 16, 20, 35
organ-pipe, note of, 17

parabola, 13
path-curve, 13
pedestrian, 13
perihelion of Mercury, 93–94
physical reality, 101, 106, 109
physical theory
    limiting case, 58
physics, 12
    of measurement, 12
place specification, 11
plane, 81
Poincaré, 81
point-mass, energy of, 36
ponderable body, 106
ponderable mass, 106
ponderable matte, 149
ponderable matter, 106, 107, 131
position, 13
Pound and Rebka experiment,
    130
Principe, 96
principle of equivalence, 110, 111
principle of inertia, 109
principle of relativity, 16–17, 20,
    47
principle of the conservation of
    mass, 129
principle of the constancy of the
    velocity of light, 109,
    128
principle-theories|hyperpage,
    127, 158
processes of Nature, 34
propagation of light, 19, 20, 27,
    68, 89, 137, 158
    in gravitational fields, 57
    in liquid, 32
proposition

"true", 9
correct, 9
psychological origin of the
    concept of time, 103
psychological origin of the idea
    of space, 101
Pythagorean theorem, 111

quasi-Euclidean universe, 84
quasi-spherical universe, 84

radiation, 36
radioactive substances, 39
real external world, 103
reference-body, 11–15, 19, 22–24,
    30, 40, 47, 52, 72
    rotating, 60
reference-mollusk, 73–74
relative
    position, 10
    velocity, 88
relativistic mass
    experimental fact, 121
    pedagogical virus of, 118
    vigorous campaign against,
        117
rest, 17
Riemann, 64, 81, 82
Riemann condition, 111
Riemann, B., 111
rigid body, 11, 12
Rindler, W., 118
rotating disc, 60, 61, 96
rotating reference-body, 96
rotation, 61, 128

Schwarzschild, 98
seconds-clock, 30
Seeliger, 79
simultaneity, 21–24, 61
    conventional, 22, 155
    definition, 21
    relativity of, 24
size-relations, 67
Sobral, 96
solar eclipse, 58, 95
    photographs, 95

solar eclipse of 29th May, 1919,
    96
space, 13, 40, 42, 79
    closure (finiteness) of, 100
    conception of, 20
    empty, 102
    identical with extension, 101
    interval, 43
    point, 73
    three-dimensional, 91
    two-dimensional, 81
space coordinates, 42, 61, 73
space-interval, 26
space-like, 104
space-time, 72, 106, 112
    structural quality of the
        field, 112, 158
special principle of relativity,
    111, 128
special theory of relativity, 20,
    43, 110, 128
spherical
    space, 83
    surface, 81
St. John, 98
stellar universe, 79
straight line, 13, 62, 66, 81
structure of space, 99
system of coordinates, 11, 13, 15

temporal arrangement of the
    experiences, 104
terrestrial space, 17
theoretical physics, 137
theory
    truth of, 93
three-dimensional, 42
time
    conception of, 20, 40, 79
    coordinate, 42, 73, 122
    in physics, 21, 73, 91
    of an event, 22, 23
    proper, 122
time-interval, 26, 43
total eclipse of the sun, 95
Trafalgar Square, 11
trajectory, 13

understanding of inertia, 105
uniform translation, 16, 47
Universe
    elliptical, 84
    space expanse (radius) of, 84
    spherical, 84
universe
    elliptical, 83
    Euclidean, 81, 82
    spherical, 82
Universe (World) structure of,
    81, 84
universe (world) structure of
    circumference of, 82

value of π, 61, 82
velocity of light, 14, 19, 58, 89,
    155
    not constant, 58, 70, 155,
    156
Venus, 94

wave-field, 106
weight (heaviness), 51
Weyl, H., 137
world, 42, 82, 91
world-point, 91
world-radius, 83
world-sphere, 82
worldline
    geodesic, 123
    timelike, 122
worldtube, 123

Zeeman, 32

æther-idea, 41

"inertial system", 105
"material points", 105
"physical reality", 106
"topological space", 112

www.ingramcontent.com/pod-product-compliance
Lightning Source LLC
Chambersburg PA
CBHW072142090426
42739CB00013B/3256